Textile Progress

2008   Vol 40   No 1

Microfibres

S. Mukhopadhyay
G. Ramakrishnan

The Textile Institute

CRC Press
Taylor & Francis Group
Boca Raton London New York

CRC Press is an imprint of the
Taylor & Francis Group, an **informa** business
A TAYLOR & FRANCIS BOOK

First published 2008 by Taylor & Francis

Published 2018 by CRC Press
Taylor & Francis Group
6000 Broken Sound Parkway NW, Suite 300
Boca Raton, FL 33487-2742

ISBN-13: 978-0-415-47450-4 (pbk)

Visit the Taylor & Francis Web site at
http://www.taylorandfrancis.com

and the CRC Press Web site at
http://www.crcpress.com

CONTENTS

1. **Introduction**   1

2. **Manufacturing methods for microfibres**   2
    2.1.   Continuous filament type   2
           2.1.1   Direct spinning   3
                    2.1.1.1 Developments of microfibre production through direct
                             spinning   4
                    2.1.1.2 Advantages   4
                    2.1.1.3   Disadvantages   5
                    2.1.1.4 Microfibres through post-spinning operations after
                             direct spinning   5
    2.2.   Macromolecular characterisation of fibres through direct spinning technique 11
           2.2.1   Conjugate spinning   15
                    2.2.1.1 Islands-in-sea type   15
                    2.2.1.2 Split spinning   16
                    2.2.1.3 Development in bicomponent spinning   19
                    2.2.1.4 Multi-layer spinning   21
    2.3.   Random (staple) type   22
           2.3.1   Melt blowing   23
           2.3.2   Spun bonding   24
           2.3.3   Flash spinning   24
                    2.3.3.1 Developments in flash spinning technique   27
           2.3.4   Polymer-blend spinning   28
           2.3.5   Electrospinning   28
           2.3.6   Other random processes   29

3. **Microfibres from different polymers**   30
    3.1.   Acrylic   30
    3.2.   Polypropylene   31
    3.3.   Cellulose   31

4. **Developments in microfibre manufacturing**   32
    4.1.   Changing the cross section without changing the spinneret   32
    4.2.   Innovative way to make nonwovens from cellulosic microfibres   32
    4.3.   Radial quenching system   33
    4.4.   A new self-suction cooling device   33

5. **Texturing machines for microfibres**   33

6. **Mechanical processing of microfibres**   33
    6.1.   Introduction   33
    6.2.   Properties of microfibres as a function of diameter   34

6.3.	Spinning of microfibres		38
	6.3.1	*Blow room*	38
	6.3.2	*Carding*	40
	6.3.3	*Draw frame*	45
	6.3.4	*Roving frame*	49
	6.3.5	*Ring frame*	49
	6.3.6	*Open-end rotor spinning*	49
		6.3.6.1 *Requirements of feed sliver to adopt in rotor spinning of microfibres*	50
		6.3.6.2 *Effect of opening roll*	50
		6.3.6.3 *The twist index*	51
		6.3.6.4 *Rotor and delivery speeds*	51
		6.3.6.5 *Polyester microfibres in blends with cotton*	51
	6.3.7	*Spinning microfibre yarns on the Murata jet system*	52
	6.3.8	*Air-jet spinning of microfibres*	53
	6.3.9	*Compact spinning of microfibres*	53
6.4.	Typical processing line for a synthetic fibre		54
6.5.	Migration behaviour of microfibre in short staple spun yarns		56
6.6.	Warping		57
6.7.	Sizing		58
6.8.	Weaving		58
6.9.	Knitting		59
	6.9.1	*Properties of knitted fabrics made from microdenier fibres*	63
7.	**Techniques to improve wetting of microfibres: hydrolysis**		63
7.1.	Alkaline hydrolysis		64
7.2.	Enzymatic hydrolysis		65
7.3.	The effect of chemical splitting		66
8.	**Microfibre dyeing**		66
8.1.	General dyeing of microfibres		66
8.2.	Special considerations for microfibre dyeing		67
	8.2.1	*Reduced depth of shade*	67
	8.2.2	*Stagnant solution layer*	68
	8.2.3	*Thermomigration*	69
9.	**Various uses of microfibres**		71
9.1.	Mechanism of cleaning by microfibres		71
9.2.	Mechanism of filtration by microfibres		72
9.3.	Industrial		73
9.4.	Civil		75
9.5.	Medical		75
	9.5.1	*Polypropylene*	76
	9.5.2	*Polyethylene*	76
9.6.	Apparel		76
9.7.	Synthetic leather		77
9.8.	Household		78
9.9.	Miscellaneous		79

10. **Economics of Microfibre Processing**   80

11. **Limitations and precautions**   80

12. **Conclusion and suggestions for further work**   81

Textile Progress
Vol. 40, No. 1, 2008, 1–86

Microfibres

S. Mukhopadhyay[a,*] and G. Ramakrishnan[b]

[a]*Department of Textile Engineering, Anuradha Engineering College, Chikhli, Maharashtra, India;* [b]*KCT-TIFAC Core in Textile Technology and Machinery, Department of Textile Technology, Kumaraguru College, Coimbatore, Tamil Nadu, India*

Microfibres denote synthetic fibres that are finer than any fibre in nature. Microfibres are usually made of polyester, polyamide, acrylic, modal, lyocell and viscose in the range of 0.5–1.2 dtex. The progress starts with direct spinning and post-spinning developments for manufacturing microfibres. Researches on conjugate spinning techniques are reported along with the development in bicomponent spinning. Interesting developments in manufacturing techniques like the change of cross section without altering the spinneret, radial quenching system, etc., have been discussed. Recent developments like electrospinning have also been taken up. The mechanical processing section commences with the properties of microfibres affecting the downstream process and then discusses the processing of microfibres in blow room, carding, draw frame, speed frame and ring frame. Alternative spinning technologies like open-end, air-jet and compact spinning are dealt with. In the fabric forming systems, weaving and knitting with microfibres are discussed in depth highlighting research on such fabrics. High-speed weaving of microfibres is discussed with reference to three major technologies of projectile, rapier and air-jet weaving. The reactions of microfibres to different hydrolysis environments like alkaline, acidic and enzymatic are taken up. Dyeing of microfibres and the specific problems in dyeing of microfibres are discussed. The study of fibre structure by critical dissolution time is addressed. Different uses of microfibres in terms of industrial, medical, apparel and miscellaneous applications are presented. The economics of production along with the limitations and precautions of the fibre are subsequently discussed followed by suggestions for future work.

Keywords: microfibres; advanced spinning; microdenier fibres

1. Introduction

One of the most significant developments in recent years has been the technology to extrude extremely fine filaments while maintaining all of the strength, uniformity and processing characteristics expected by textile manufacturers and consumers. These microfibres are even finer than luxury natural fibres such as silk. This comparison, coupled with their exceptional performance, has led some in the industry to refer to microfibres as 'supernatural'.

Until now, there has been no generally accepted definition for microfibres. But, normally, the term is linked to the fibre diameter and/or weight/length of the filament in dtex or denier and not with any properties of the fibre. Fibres in the 0.1–1.0 dtex range are termed as microfibres.

Microfibres, used alone or in blends, have created considerable interest in the apparel industry because of their potentially greater comfort and functionality. Additionally, their lower diameter, greater surface area and flexibility offer many applications in areas

*Corresponding author. Email: samrat.mukhopadhyay@gmail.com, samrat@det.uminho.pt

ISSN 0040-5167 print/ISSN 1754-2278 online
DOI: 10.1080/00405160801942585

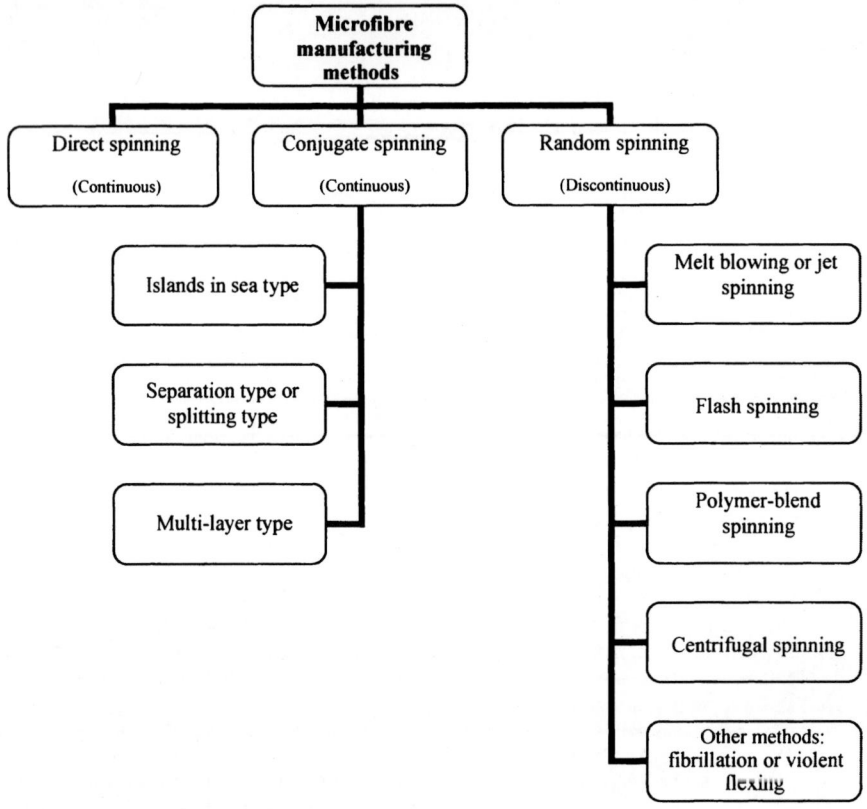

Figure 1. Classification of microfibre-manufacturing methods.

of nonwovens such as filtration, synthetic leather, protective clothing and wipes. Today, synthetic microfibres are used with polymers like polyester, nylon and acrylic.

Dr. Miyoshi Okamoto, a chemist in the Toray Industries textile research laboratory, introduced the first microfibre in the mid-1960s. Initially, the microfibre had various limitations, which were overcome by the success of *imitation leather*. In many product lines, it is the luxurious feel and look of the fabrics that makes microfibres so special. In others, it is its unique physical and mechanical performance. Today, the microfibres have taken a place of their own by the virtue of their unique properties and end uses.

Microfibres are classified into two types: (1) a continuous filament type and (2) a random (staple) type. The continuous type has again two broad divisions—direct and conjugate spinning. The detailed classification is given in Figure 1.

There have been developments in both the fields of microfibre manufacturing. Some of the techniques in random spinning, however, have been very specialised and have not seen many commercial applications.

2. Manufacturing methods for microfibres

2.1 Continuous filament type

There are two ways to produce microfibres of the continuous filament type – direct spinning and conjugate spinning. Direct spinning and post-spinning processes were earlier to develop followed by conjugate spinning.

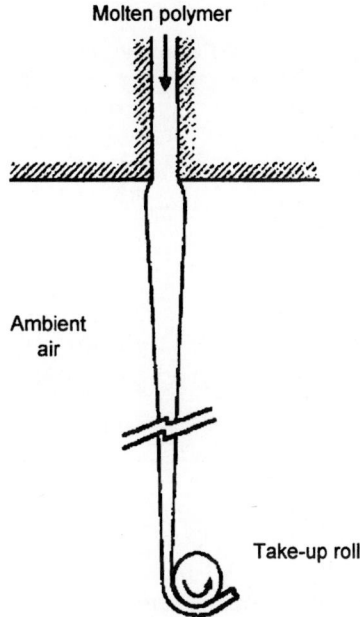

Figure 2. Melt spinning.

2.1.1 Direct spinning

In conventional spinning, the polymer is ejected into a gas or solution and is subsequently drawn. The polymer is either melted (melt spinning) or converted into a solution (solution spinning) before it can be extruded through fine holes called spinneret. For microfibres, however, the majority of the developments have happened in the field of melt spinning (Figure 2). The direct spinning method can be visualised as an extension to conventional spinning. In the application of conventional melt spinning, there are problems like the following:

(1) Melt fracture (Fibre breakdown) (dripping).
(2) Variation of filament thickness.
(3) Spinneret clogging.
(4) Denier variability among filaments in a single yarn.

   Since microfibres have dimensions much finer than conventional fibres, the following precautions are taken to avoid these problems, as summarised by Kasei et al. [1].

(1) Optimisation of polymer viscosity (i.e. a higher spinning temperature to reduce viscosity)
(2) Optimisation of the spinneret design (i.e. the spinning holes arranged to ensure homogeneous cooling)
(3) Optimisation of the ambient temperature underneath the spinneret (i.e. quenching, cooling rate control)
(4) Optimisation of filament assembly (i.e. assembly nearer to the spinneret)
(5) Optimisation of spinning draft (i.e. spinning tension control)
(6) Lower rates of extrusion (i.e. stable polymer transmission)
(7) Purification of spinning polymer (i.e. high-efficiency filtration)

Table 1.   Spinning Conditions for PET Ultra-Fine Fibre by Direct Spinning [148].

Spinning conditions	<0.3 d	<0.15 d
Number of holes in spinneret	>140	>150
Cross section of output hole (cm)	<3.5 × 10⁻⁴	<1.5 × 10⁻⁴
Polymer melt viscosity (poise)	<950	<300
Ambient temperature at a point 1–3 cm below spinneret (°C)	<200	<150
Filament-assembling position (cm below spinneret)	10–20	20–70
Drawing	Conventional	Conventional
Tenacity (g/d)	3–5	3–5
Elongation (%)	20–40	20–40

2.1.1.1. Developments of microfibre production through direct spinning. The Unitika Co. was the first to put microfibres of 0.3–0.5 denier in the market, although Asahi Chemical Industry Co. developed finer polyester fibre of 0.1–0.3 denier by optimising the polymer melt viscosity, the spinneret design, the ambient temperature underneath the spinneret (i.e. the cooling conditions) and the filament assembly conditions. Unitika adjusted the polymer melt viscosity to less than 950 poise, the cross-sectional area per spinneret hole to less than 3.5×10^{-4} cm², and also controlled the ambient temperature at 1–3 cm underneath the spinneret below 200°C. The extruded filaments were assembled at 10–200 cm underneath the spinneret.

Asahi Chemical Industry Co. succeeded in producing micropolyester fibre of less than 0.15 denier by extruding the polyester of melt viscosity less than 480 poise through a spinneret of above 300 holes. Each hole was of less than 1×10^{-4} cm² cross-sectional area and arranged concentrically. However, the extruded polymer formed droplets and exhibited no drawability unless the thermal environment immediately below the spinneret was suitably controlled. The ambient temperature at 1–3 cm underneath the spinneret holes must be kept below 152°C by blowing cold air from the circumference of the spinning threadline to enable the polymer to be drawn into filaments. These concentrically arranged filaments should all be cooled at the same rate. Then the filaments need to be assembled at 20–70 cm underneath the spinneret holes, and wound up as underdrawn fibre. This underdrawn fibre can be drawn conventionally to yield ultra-fine fibre less than 0.15 denier. Table 1 summarises the spinning conditions for ultra-fine polyester fibre production by direct spinning.

Kasei and Co. investigated the influence of air friction on the high-speed spinning of ultra-fine fibre [1]. Toray and Toyobo have also produced ultra-fine fibre by direct spinning. Unitika has succeeded in producing cationic-dyeable ultra-fine polyester filaments by using homogeneous cooling temperatures at spinning.

Overall, the direct spinning process has some advantages and suffers from some limitations. They are summarised as follows:

2.1.1.2. Advantages.

(1) The process of direct spinning is comparatively simple and easy to control.
(2) A single-component ultra-fine fibre is obtained by direct spinning, and the subsequent processes require no complicated processing such as splitting into two components or removing a second component.

2.1.1.3. Disadvantages.

(1) The direct spinning of microfibres is a complex process and involves the modification of the spinning methods used to produce conventional denier fibres.
(2) Generally, there is a limit to the fineness of fibres produced using direct spinning, as exemplified by polyester that cannot be extruded at less than about 0.15 g/min because of breakage of the filaments. Typically, microfibres in the range of 0.2–0.9 denier are secured using conventional spinning methods, although filaments of 0.1 denier have been reported.
(3) Filament breakage and fluff formation are often observed in direct spinning of ultra-fine fibres, and a handle of high-quality cannot be expected.
(4) As with conventional direct spinning, the microfibre manufacturing process suffers from problems like fibre breakdown, variation in filament thickness and spinneret clogging.

2.1.1.4. Microfibres through post-spinning operations after direct spinning.

Laser heating

Suzuki and colleagues have tried a different method for producing microfibres that involved spinning the fibre and drawing the fibre through laser heating. They have experimented the technique for all fibre-forming polymers available and also report a work where they have used it for nonwovens.

Polyester microfibres were produced by Suzuki and Mochizuki [3], using the laser-heating technique. First, the original poly(ethylene terephthalate) (PET) fibre was zone drawn at a drawing temperature of 90°C under an applied tension of 4.9 MPa, then the zone-drawn fibre was zone-annealed at 150°C under 50.9 MPa.

The CO_2 laser-heating apparatus used for producing the microfibre consisted of the continuous-wave CO_2 laser emitter (PIN10S; Onizuka Glass Co., Ltd., Japan), a power metre with a thermal head and an electric slider (Limo Oriental Motor Co. Ltd., Japan), as shown in Figure 3. The electric slider was used to move the fibre at a constant speed. The continuous-wave CO_2 laser emitted light at 10.6 m, and the laser beam was a 4.0-mm-diameter spot. A power density was measured by the power meter before the laser heating. One end of the zone-annealed fibre was connected to a jaw equipped with the electric slider, whereas the other was attached to a slight weight. The zone-annealed fibre, moving downward at a speed of 500 mm/min, was heated by irradiating the continuous-wave CO_2 laser and drawn instantaneously. The apparatus used in the zone drawing and zone annealing was previously described in detail elsewhere [11].

The thinning mechanism as explained by the authors was as follows. The higher crystallinity and crystal perfection in the annealed fibre make it possible to irradiate a higher output laser to the fibre as compared with the thinning for the original fibre, the plasticity viscosity in part heated by a high-output laser becomes extremely low. However, the fibre was not stretched or broken because of an extremely low applied tension, and the heated part becomes nearly molten, that is, a sintering state.

The microfibre was spun from the sintering state. This is clear from the microphotograph of a spindle-shaped necking as shown in Figure 4.

The same method has been tested by the researchers for polypropylene [4], nylon 6 [5] and nylon 66 [6] also.

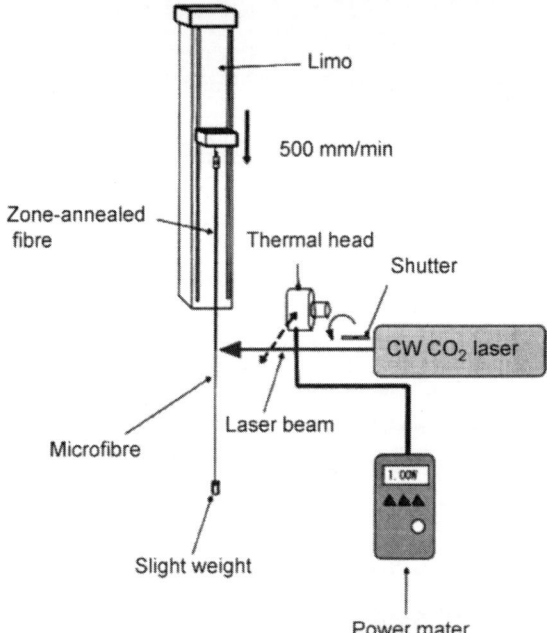

Figure 3. Apparatus used for laser heating (discontinuous type) (Reproduced with permission from Suzuki and Mochizuki [3]).

Suzuki and colleagues later reported an apparatus that can continuously wind up microfibre (Figure 5) as a monofilament in the winding speed range of 100–2500 m/min. Suzuki and Narusue [7] also reported that the degree of perfection of crystallites improves for the microfibres with increasing winding speeds, as has been demonstrated by Figure 6 in the differential scanning calorimetry (DSC) curves.

Suzuki and Tojyo [8] tried the process for poly(ethylene-2,6-naphthalate) (PEN) fibres. Figure 7 shows the SEM micrographs of fibres wound at different winding speeds showing the reduction of diameter of the fibres and the subsequent increase of birefringence (Figure 8). The increase of the birefringence was caused mainly by increasing the degree of amorphous orientation because no reflection due to the crystallites was observed in the WAXD patterns.

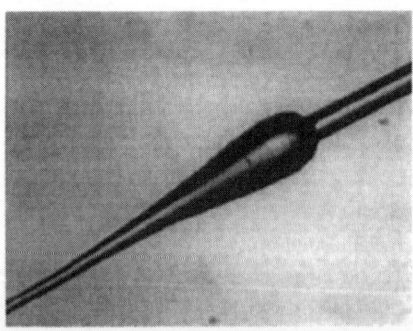

Figure 4. Microphotograph of a spindle-shaped necking in the laser-heated fibre (reproduced with permission from Suzuki and Mochizuki [3]).

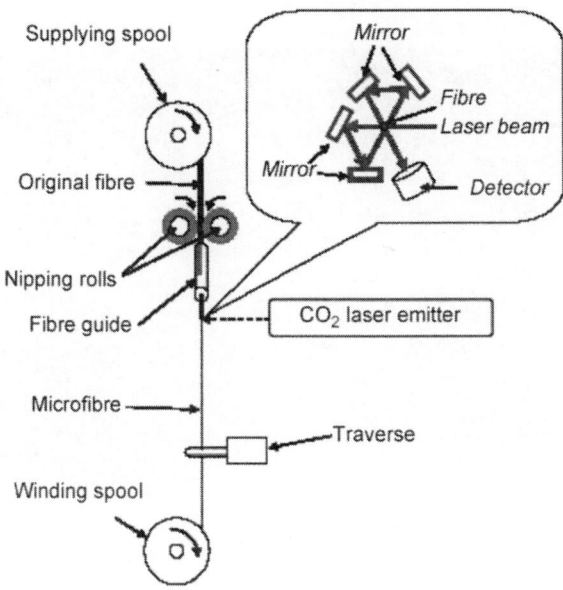

Figure 5. The continuous method of laser heating (Reproduced with permission from Suzuki and Mochizuki [3]).

The authors, as explained in the previous paragraphs, had manufactured nylon 6, nylon 66, polypropylene and poly(L-lactic acid) that were obtained by the laser-thinning; the authors also had the oriented crystallites grown by the flow-induced crystallisation; and the degrees of crystal orientation and crystallinity increased as the S_w increased. The

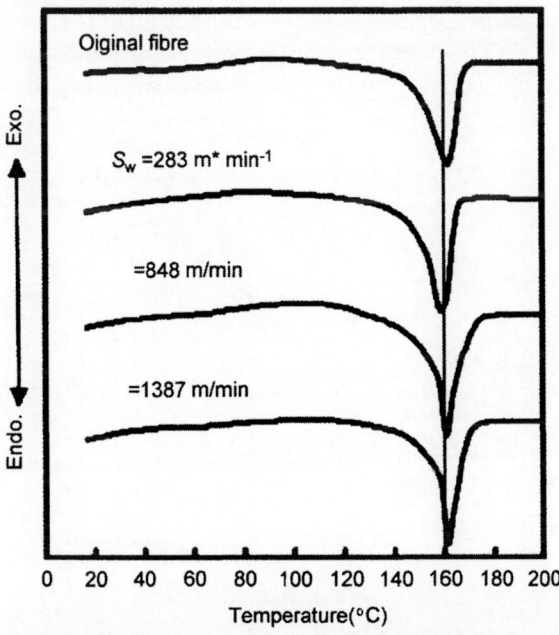

Figure 6. DSC curves of the original fibre and the isotactic polypropylene (i-PP) microfibres wound at three different winding speeds (Reproduced with permission from Suzuki and Narusue [7]).

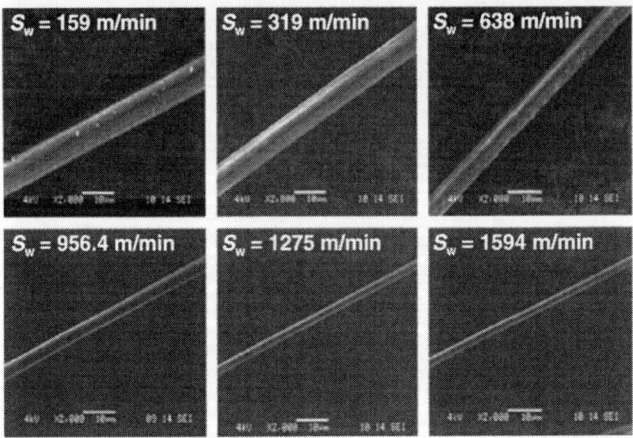

Figure 7. SEM photographs for the microfibres wound up at a series of winding speeds (Reproduced with permission from Suzuki and Tojyo [8]).

observation has been different for PEN microfibre, where an increase in overall crystallinity was observed without an increase in the oriented crystallites. The authors explained this observation as a result of the absence of flow-induced crystallisation that did not occur because of the existence of naphthalene ring in the main chain structure.

Isotactic polypropylene hollow microfibre (Figure 9) was continuously produced by the group [9] by using a CO_2 laser-thinning method. To prepare the hollow microfibre continuously, the apparatus used for the thinning of the solid fibre was improved so that the laser can circularly irradiate to the hollow fibre.

In laser thinning, the temperature of the fibre irradiating the CO_2 laser beam instantly reaches near its melting temperature (T_m), and the melt viscosity in the part of the fibre heated becomes sufficiently low without elongation or cutting, and the heated part enters a nearly

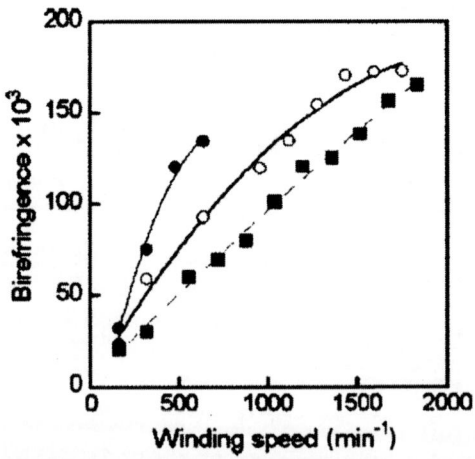

Figure 8. Winding speed dependence of the birefringence of PEN microfibres obtained at various supplying speeds (Ss), (●) Ss = 0.16, (○) Ss = 0.33, and (■) Ss = 0.63 m/min (Reproduced with permission from Suzuki and Tojyo [8]).

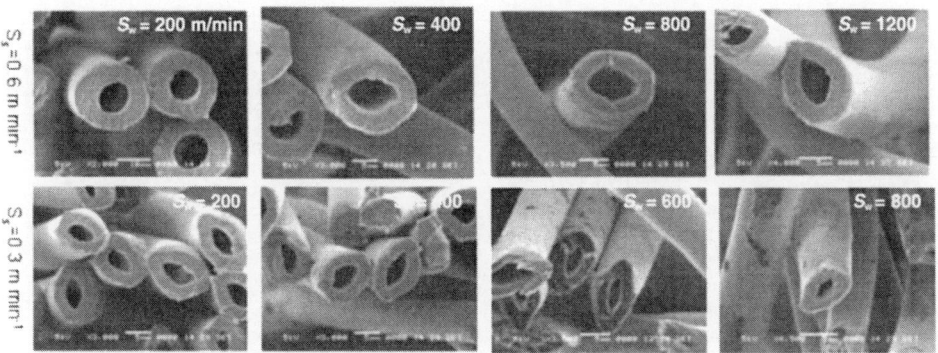

Figure 9. SEM photograph of i-PP hollow microfibres obtained at various winding speeds (Sw) at supplying speeds (Ss) of 0.3 and 0.6 m/min (after Suzuki and Ohnishi [9]).

molten state. The fibre is continuously thinned (Figure 10) by the plastic flow occurring from the nearly molten state, and then the microfibre, having fallen down because of its own weight, is wound on the winding spool. The instantaneous plastic flow at a high strain rate induces the molecular orientation and crystallisation despite a large deformation, just like in flow drawing, and gives an oriented microfibre. The laser thinning, which occurrs as an instantaneous large deformation, differs largely from conventional drawings and melt spinnings in deformation process.

The hollowness of the fibres, as investigated by the Suzuki and colleagues, decreased slightly when compared with the original one. In spite of the laser thinning due to the

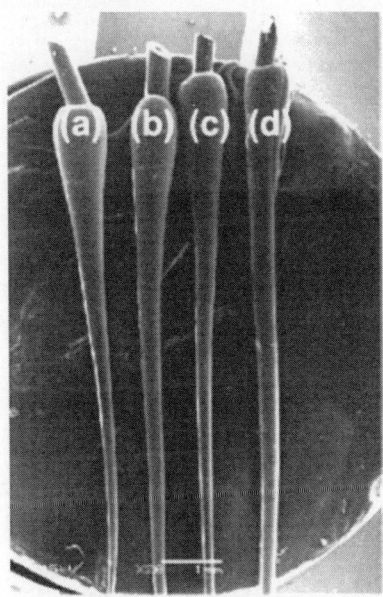

Figure 10. SEM photographs of the necks of hollow fibres obtained at four different winding speeds (Sw): (a) $S_w^{1/4}$ 1200 m/min, (b) $S_w^{1/4}$ 800 m/min, (c) $S_w^{1/4}$ 400 m/min, (d) $S_w^{1/4}$ 200 m/min (reproduced with permission from Suzuki and Ohnishi [9]).

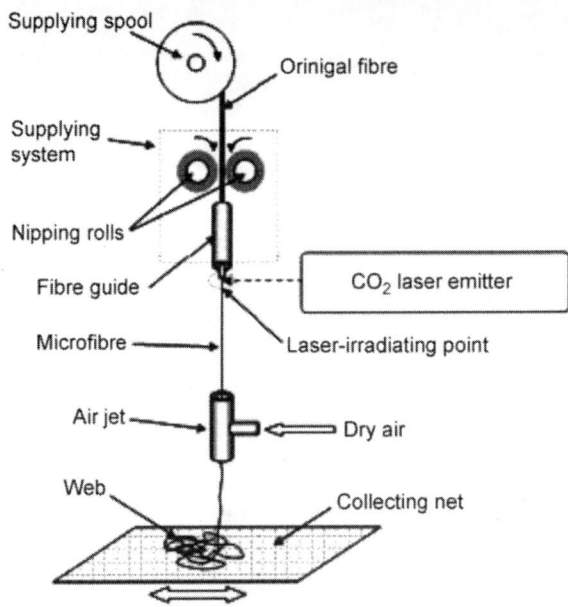

Figure 11.   Laser-thinning method used to prepare nonwoven material (after Suzuki and Kishi [10]).

partial melting induced by the laser irradiation, the hollowness was intact, and such hollow structure will not be able to be made by any conventional spinning. As the microfibres are fit for various purposes, many applications of the hollow microfibre will be found in medical and industrial fields.

PET nonwoven fabric was prepared by Suzuki and Kishi [10] from microfibres (Figure 11) that were previously produced using a laser-thinning method. The PET nonwoven fabric obtained was made of the endless mircofibres with a uniform diameter without droplets. The fibre diameter was varied by controlling airflow rate into the air jet and supplying speed of an original fibre into a laser-irradiating point. The obtained microfibre was slightly oriented along fibre axis by the flow-induced crystallisation.

Figure 12 shows the SEM photographs, at a magnification of 1500, for the webs obtained for four airflow rates. The four webs were obtained by laser-irradiating the original fibre supplied at $S_s^{1/4}$ 0.15 m/min. The observation of the SEM photographs shows that the webs have a smooth surface without a surface roughened by laser ablation, and there is no droplet in the webs.

Further treatment of fibres formed through laser heating

Suzuki and Mizuochi [11] improved the tensile properties of poly-L-lactic acid (PLLA) fibres. The fibre obtained by the method of laser heating was subsequently treated through the zone drawing-annealing process. It is difficult to draw and anneal only a microfibre obtained, since its diameter and tensile strength are too small, but the microfibres in a bundle could be drawn and annealed by zone heating. To zone draw and zone anneal the microfibre bundle, an apparatus was specially constructed as shown in Figure 13. This apparatus consists of a temperature-controlled zone heater and a Linead motor (Oriental Motor Co. Ltd., USA) capable of moving the zone heater at an arbitrary speed. One end of the microfibre bundle was connected to a fixed support while the other was connected to a

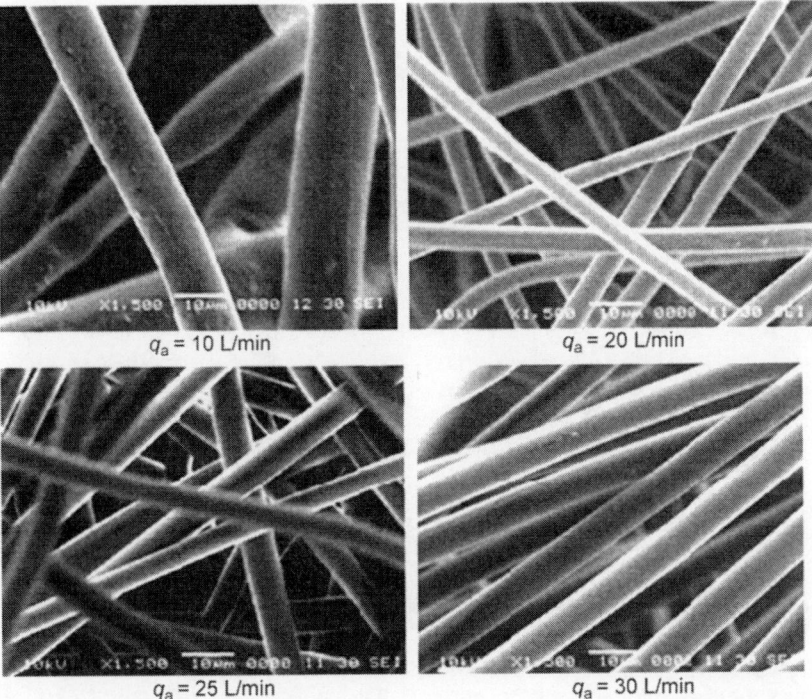

$q_a = 10$ L/min $q_a = 20$ L/min

$q_a = 25$ L/min $q_a = 30$ L/min

Figure 12. SEM photographs of 1500 magnifications for the microfibres obtained at four different airflow rates (after Suzuki and Kishi [10]).

weight after passing through a pulley. The drawing and annealing treatments were carried out by moving the zone heater along the fibre direction under an optimum tension.

2.2 Macromolecular characterisation of fibres through direct spinning technique

The literature on microfibre characterisation is scanty. In one of the detailed investigations by Die'val et al. [12], they investigated three different fibres—two of normal denier but varying cross section and a microfibre. The crystallinity index (CI) was determined through WAXD by isolating the crystalline peaks from the total signal. The microfilament,

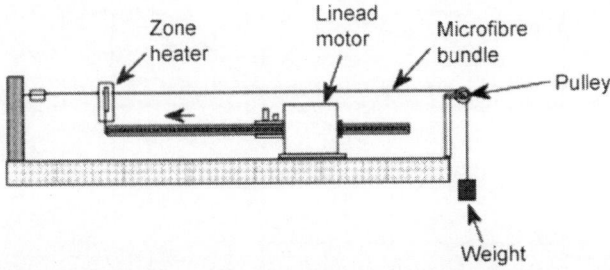

Figure 13. Schematic diagram of apparatus used for zone drawing and zone annealing of microfibre bundle (after Suzuki and Kishi [10]).

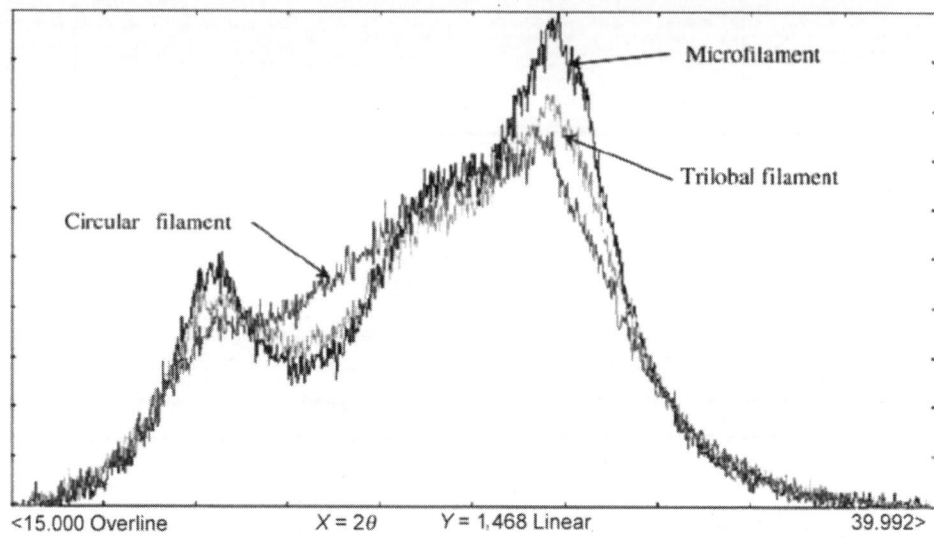

Figure 14. X-ray diffraction figure for the three kinds of fibres (after Die'val et al. [12]).

which was a direct spun and subsequently drawn filament, had the highest CI of 43.3%
(Figure 14).

The evolution of CI (Figure 15) with temperature was also interesting. At annealing
temperature ranging between 100°C and 120°C, the CI was stable for the circular fibre.
The CIs of the trilobal fibre and of the microfibre increased slightly between 100°C and
110°C then remained stable between 110°C and 120°C. Their CIs remained close during
this evolution. After a heat treatment at 135°C, the CI of each fibre attained the same

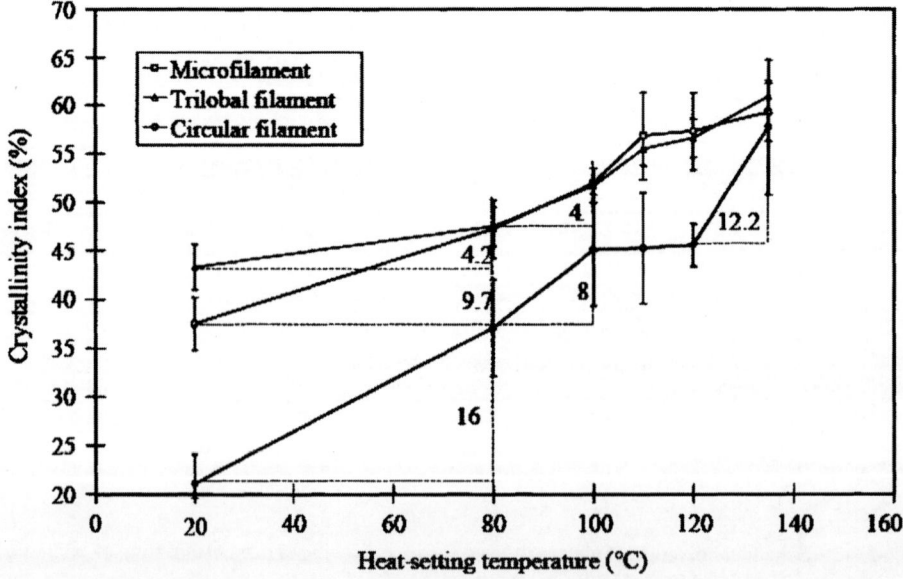

Figure 15. Evolution of crystallinity index. (After Die'val et al. [12])

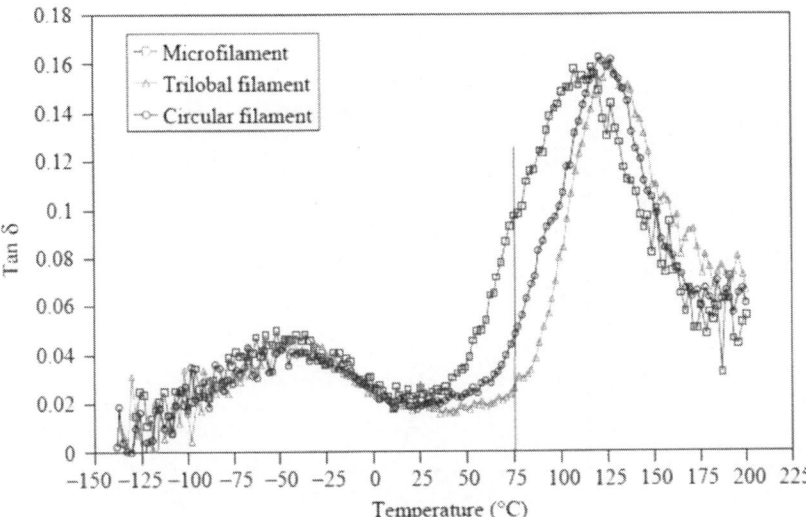

Figure 16. Evolution of loss-angle tangent at a test temperature of 75°C for the three kinds of fibres (after Die'val et al. [12]).

value, close to 55%. To reach this value, the CI of the circular fibre increased strongly (increase, 21%). On the other hand, the CIs of the other two fibres practically did not increase between 120°C and 135°C. The authors, however, did not give any strong reasons behind such observations.

Dynamic mechanical studies also highlighted some important results. The dynamic modulus, characterised by Visco-analyzer VA 815 from Metravib Instruments (Lyon, France) has two components. The first is called E_0 (storage modulus) and characterises the elastic field of the material. The second is the loss modulus E_{00}. This modulus corresponds to viscous dissipations. In the range of temperatures used, the evolution of the modulus E_0 showed that the glass transition temperatures were close to 100°C. The glass transition is very important from the purpose of dyeing also. During this transition, the macromolecular chains became more mobile, resulting in a drop in the elastic modulus.

Figure 16 shows the behaviour of all three fibres before heat treatment. The glass transition was located at around 110°C. The β-transition was located towards 250°C. But the peak of the corresponding loss-angle tangent does not have a very significant amplitude and does not allow easy discrimination between fibres. The identity of each fibre is well highlighted by observing the value of the tangent of the loss angle at a given temperature (75°C, as has been done by the authors). At this temperature, the loss angle of the microfibres was practically equal to half its maximum value. On the other hand, for the other fibres, their values remained lower than the maximum of the β-transition. This was a sign of higher molecular mobility in microfibres, as argued by the authors. It was interesting to note that microfibres presented a progressive increase in the maximum temperature of the loss-angle tangent. This evolution was less regular for the circular and trilobal fibres.

As commented by the authors, manufacturing sequences have a very significant influence on macromolecular organisation of the fibres. This is, in turn, reflected in dyeing behaviour. A filament with mobile molecular segments will more easily accept dye molecules. The diffusion of the dyes into the filaments will also be faster. The study revealed that to

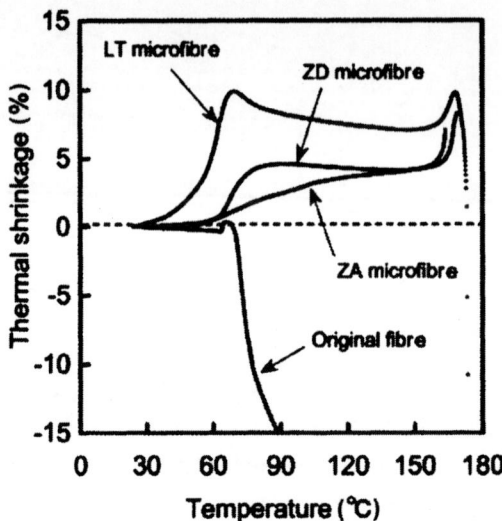

Figure 17. Temperature dependence of the thermal shrinkage on the original fibre, LT, ZD and ZA microfibres (after Suzuki and Kishi [10]).

limit its evolution of the macromolecular structure, it is necessary to maintain the orientation of the macromolecules by the action of an external constraint during annealing.

Figure 17 shows the temperature dependence of the thermal shrinkage on the original fibre as well as the microfibres obtained by laser thinning (LT), zone drawing (ZD) and zone annealing (ZA), as already reported in the work by Suzuki and Kishi [10] on PLLA fibres. The development of the thermal shrinkage during heating is associated with the chain coiling in the oriented amorphous regions and is dependent on draw ratio and the degree of crystallinity. The original fibre stretches rapidly above 67°C, and the stretch exceeds the instrumental limitation after a slight thermal shrinkage occurred at 64°C.

The rapid stretch of the original fibre shows that no physical network that was built up by the crystallites preventing the fluid-like deformation exists, and that no strain-induced crystallisation occurs during the measurement.

The orientation, however, had significantly improved as confirmed from Figure 18, which shows the WAXD patterns of the LT, ZD and ZA microfibres. The equatorial

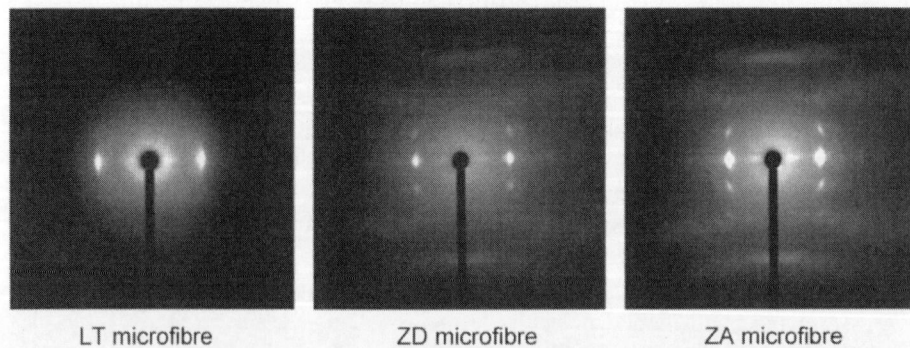

| LT microfibre | ZD microfibre | ZA microfibre |

Figure 18. WAXD patterns of the LT, ZD and ZA microfibres (after Suzuki and Kishi [10]).

reflection due to oriented crystallites was observed in the WAXD patterns. The WAXD patterns of the ZD and ZA microfibres showed strong equatorial reflections due to the highly oriented crystallites, and the strong equatorial reflections are attributable to the $(0010)\alpha$ reflection of α form crystal.

The LT microfibre shrinks rapidly in the temperature range of 50°C–70°C, stretches gradually in the temperature range of 70°C–160°C, shrinks rapidly near the T_m and then shrinks rapidly. The peak of thermal shrinkage near the T_m shows that the physical network was broken and that the fluid-like deformation occurred. The ZD microfibre stretches gradually with increase in temperature after the rapid shrinkage occurred in the temperature range of 65°C–75°C, and then the rapid shrinkage takes place near the T_m. The ZA microfibre shrinks gradually as the temperature increases and shrinks rapidly within a narrow range of temperature around the T_m. The difference in the behaviour of the shrinkage between ZD and ZA microfibres depends mainly on the molecular mobility of the amorphous regions. The crosslink density of the physical network built up by the crystallites was thought as influencing the molecular mobility in the amorphous region.

The ZA microfibre finally obtained had a diameter of 1.2 μm, a birefringence of 43.3 $\times 10^{-3}$, a tensile modulus of 14 GPa and a tensile strength of 1.5 GPa. The zone drawing and zone annealing methods were found to be effective in producing the PLLA microfibre with high modulus and high strength.

2.2.1 Conjugate spinning

Conjugate spinning is another popular way in which various innovations have taken place in bicomponent spinning of yarns. In this spinning process, the islands-in-sea type, the split type and the multi-layer type are the major varieties. Before going to the details of the individual processes, the development in the spinneret design for bicomponent spinning would be relevant. The different spinning techniques in bicomponent spinning are discussed in the paragraphs below.

2.2.1.1. Islands-in-sea type. In 'sea–island' or 'islands-in-sea' type fibres, one portion is in a dispersed phase (called island component) and the other is in a continuous phase (sea component). It is made by the conjugate spinning with two components immiscible in thermodynamics.

In the manufacturing process, one polymer is fed in individual streams inside a sea of another polymer. The spun filaments have a total denier of 2–5 dpf (12–20 μm) after drawing. The sea polymer is dissolved away generally after the fibres are knit or woven into fabric to leave small submicron island filaments on the surface of the fabric. Twenty-four and 32 islands-in-sea fibres have been around for a number of years and are used to make products such as ultra-suede and artificial leather. Now commercial operation with 64 islands for both staple and filament yarns is in practice (Figure 19).

Typical polymer ratios are 20% sea and 80% islands. When the island polymer is greater than 65% of the total filament mass, the island filaments become square in shape due to the packing density.

The microfibre fineness can be defined as follows:

$$D = \frac{dl(R/100)}{N},$$

where dl is the extruded fibre fineness; R, the island content; and N, the number of islands.

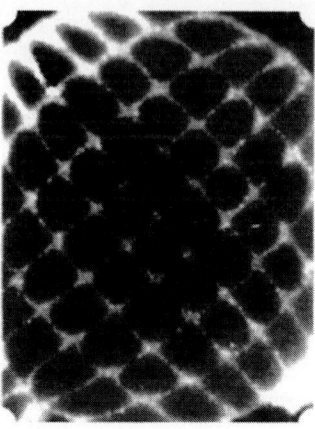

Figure 19. Cross section of an islands-in-sea fabric [13].

Hills Inc. has gone to the extent of producing 600 (Figure 20) and 900 islands (Figure 21) in sea. Islands-in-sea fibres with 1120 islands were spun on the Hills in-house pilot line in conjunction with North Carolina State University. When drawn, the final total filament denier of 1 dpf was produced, resulting in island fibres of approximately 500 nanometer diameter.

2.2.1.2. Split spinning. In split spinning, the starting fibre consists of segments of two different polymers. Each wedge of polymer A has a wedge of polymer B on either side. The general production principle of split fibres is elucidated in (Figure 22). The fibres are designed to split into the wedges by different treatments to produce the ultimate microfibre. Segmented Pie fibres are, therefore, known as 'splittable fibres'. The cross section of typical splittable fibre, as shown in Figures 23 and 24, shows a fibre after splitting.

This type of spinning distinctly differs from the islands-in-sea type in its aim to utilise the second component in the final product as well as the first by splitting the two components mechanically instead of removing the second component by dissolving. In split spinning [17], there are various methods.

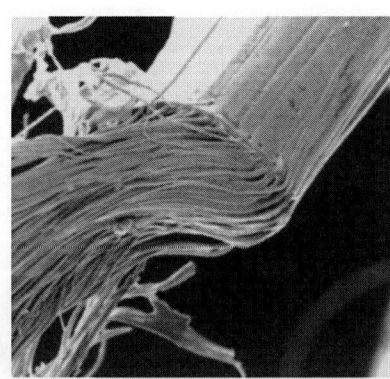

Figure 20. 600 Islands-in-sea shown at 1500× magnification [14] (www.hillsinc.com).

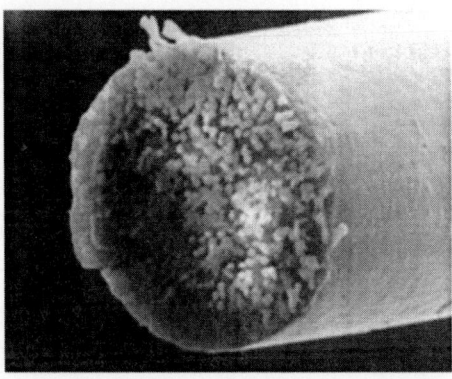

Figure 21.　900 Islands-in-sea (polyester islands in polyethylene sea) (www.hillsinc.com).

The first method includes the steps of forming fibrillisable or splittable multicomponent conjugate fibres into a fibrous structure and then treating the fibrous structure with an aqueous emulsion of benzyl alcohol or phenyl ethyl alcohol to split the composite fibres.

A second method has the steps of forming splittable conjugate filaments into a fibrous structure and then splitting the conjugate fibres of the fibrous structure by flexing or mechanically working the fibres in the dry state or in the presence of a hot aqueous solution. Hills Inc. used this concept to produce a 2–4 dpf bicomponent pie yarn, which is then spun and processed with standard techniques. In fabric form, a mild caustic solution is applied to the fabric causing the individual fibres to split apart from the main fibre. If a 32-segment pie of nylon/polyester (N/P) is used, as shown, the final dpf of the fibres are in the range of 0.1.

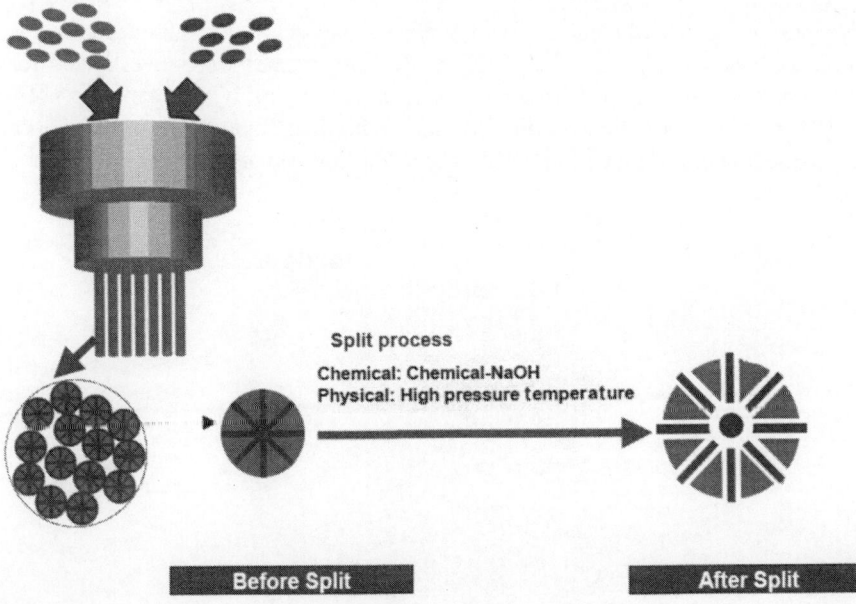

Figure 22.　Schematic diagram of the split spinning process.

Figure 23.   Cross-sectional photomicrograph of a 16-segment splittable fibre [15].

In the third method, the conjugate fibres are hydraulically or mechanically needled to fracture and separate the cross sections of conjugate fibres, forming fine-denier split fibres. The hydro-needling process [20] utilises a pressurised stream of water to split multicomponent conjugate fibres. In general, the process simultaneously splits and entangles the fibres to form a bonded nonwoven web. It is important that the hydro-needling process has not been used to produce split melt-blown fibre webs since the autogenously bonded melt-blown fibre webs, which have very fine breakable fibres and contain substantially uniformly distributed numerous inter-fibre bonds, restrict fibre movements and are difficult to split.

Carding has also been used to split fibres (Figure 25). The fibre is blended up to 30% of the fibre blend with other fibres. The fibre must be properly preblended and opened to maximise blend uniformity before carding. Worker and stripper loading must be considered when running this fibre because the split fibre has 16 times the number of fibres that must be processed through the system [19].

Various side-by-side bicomponent melt-blown fibre webs were produced on Reicofil bico melt-blown line by Sun et al. [20]. Several approaches were investigated for the subsequent fibre splitting in this research. The authors found hydroentangling not quite suitable for the bico melt-blown fibres. More fibre breakage rather than splitting occurred in the hydroentangling process, indicating that the interfacial adhesion within the bico

Figure 24.   Cross section of a microfibre after splitting [16].

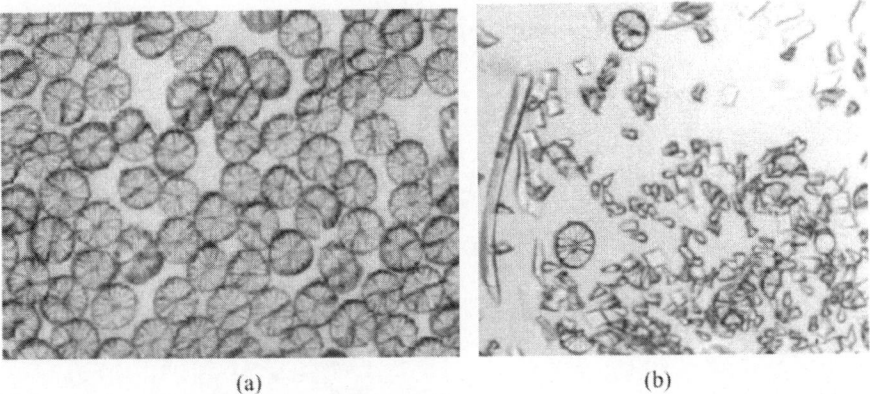

(a)   (b)

Figure 25. Fibres: (a) before splitting; (b) after splitting.

fibre is comparatively too strong than that in the weak low-molecular weight orientated melt-blown fibres. On the other hand, chemical method (alkali treatment) and hot aqueous medium inducing fibre splitting appeared suitable for the bico melt-blown fibre splitting. Benzoic acid was the most effective agent in this study for splitting PET/polyamide (PA) and polypropylene (PP)/polycaprolactam (PA6) bico melt-blown web, while no portion of the fibre was removed. Hot water treatment also led to fibre splitting in the PP/PA6 webs.

Different cross sections can be produced with the split technology. Splittable fibres can be produced with tipped trilobal cross section. Tipped trilobal bicomponent fibre is a new concept in which the second polymer is placed in a small quantity on the tip of a trilobal or delta cross-section core. After spinning, the fibres are twisted and then wet heat is applied. Therefore, the polymer on tips of the fibre breaks apart into microfibres and spiral around the core polymer [21].

Advantages

Split-type microfibre fabrics have numerous capillaries [22] due to the splitting of the conjugate filaments. These fabrics are composed of synthetic filaments that usually have hydrophobic characteristics, so they rapidly absorb and transport moisture because of these capillaries. Split-type-microfibre fabrics also have drapeability, softness, bulkiness and smoothness so there are many applications for them, such as general wiping cloths, food service, clean rooms, medical parts, etc.

Split-type N/P microfibres are widely used for artificial suede, wiping cloths, peach-skin fabrics, silk-like fabrics, and air-permeable and water-proof fabrics due to their many advantages of enhanced drapeability, softness, bulkiness, smoothness, high aesthetics and good comfort properties compared with regular polyester fibres. Moreover, splitting N/P microfibres yields fabrics with fine, closely packed and aligned capillary columns of water between the fibres and larger surface areas on the fabrics, resulting in excellent absorbency [23]. The triangular cross-sectional shape and ultra fineness of these N/P microfibres, formed by splitting composite fibres, provide the microfibre fabric with excellent wipeability due to the sharp-edged effect of split microfibres.

2.2.1.3. Development in bicomponent spinning. One of the first patents in the field of bicomponent spinning goes to Bryan et al. [24].

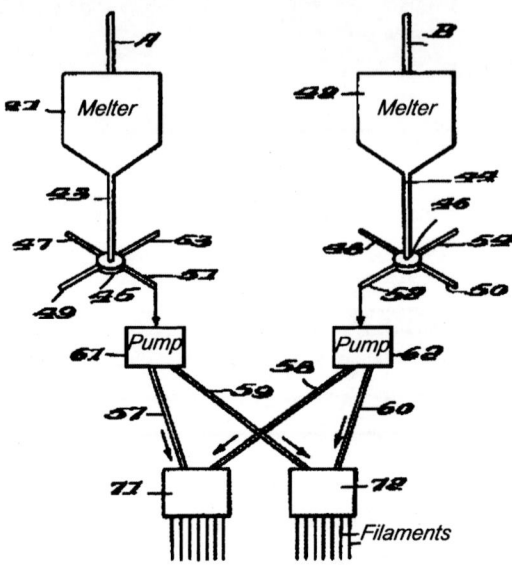

Figure 26.   Schematic diagram of bicomponent spinning (after US Pat. 3381074).

From Figure 26, the melters 41 and 42 are for two different polymers through metering pumps 61 and 68. Transfer pipes 57 and 59 transfer dual streams from the left hand pump to the spinneret assembly from where the bicomponent fibres are extruded.

US Pat. No. 3531368 [25] illustrates the details of a nozzle and describes a process for the manufacture of a matrix microfilament yarn where fine microfilaments (segments) of component A are surrounded by a matrix component B and separated from each other by the latter. This type of structure is obtained firstly by pre-molding bicomponent structures of core-skin or side-by-side structure. Subsequently, many such structures are collected in a funnel-like chamber opening into a spinning orifice and extruded through the spinning orifice.

The improvement that Okamoto incorporated in US Pat. No. 3692423 [26] was by changing the converging angle of the funnel-shaped space θ to 75° (Figure 27) Conduit 24 supplies the sea constituent polymeric liquid, whereas the island component is supplied through conduit 25. Chamber 27 distributes the island constituent polymeric liquid into the spinneret 23. The chamber 26 is fluid tight and separated from chamber 27 by annular packing 28. The converging angle of 75° is effective for gradually decreasing the diameter and uniformly uniting the composite streams in a stabilised condition.

There were several further improvements in spinneret design. One significant development was by Moriki and Ogasawara [27], who came out with a noteworthy improvement in the spinneret design. The main problem encountered with earlier spinnerets was in the control of viscosity for the islands and sea components. It was found in many cases that the island component fused with the sea component.

The improved spinneret (Figure 28) designed by Moriki and Ogasawara [27] was an enhancement of the earlier versions. The new design allowed the constituent streams, substantially engrafted in a sea constituent, to flow in such a manner that they are uniformly distributed and separated from each other in a cross-sectional view.

Figure 28 describes the spinneret designed by Moriki and Ogasawara. The design consisted of an upper horizontal plate 21 with vertical inlet nozzles 27 extending downward

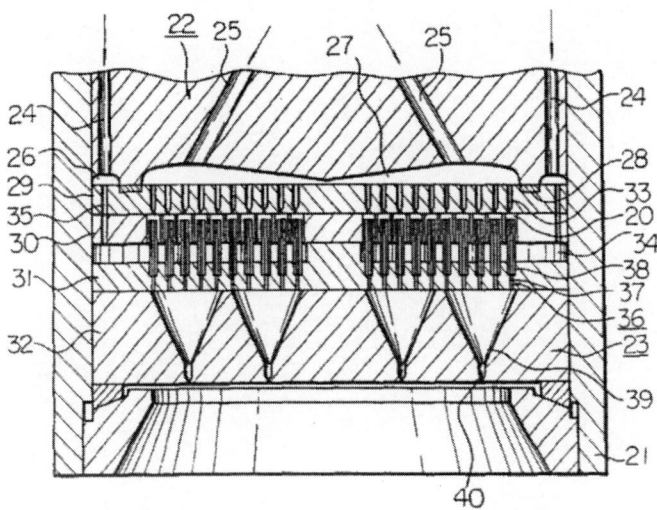

Figure 27. The spinneret assembly for islands-in-sea spinning (after US Pat. No. 3692423).

for a sea constituent melt A. Each nozzle group consists of the same number of inlet nozzles 27. A lower horizontal plate 22 has groups of vertically extending inlet holes 28. The hole groups correspond to the respective nozzle groups. Each of the inlet holes 28 is paired with a corresponding inlet nozzle 27 in such a manner that each nozzle extends into the inlet hole 28. The inlet nozzle and the inlet hole in combination form a circumferential space 26 for the passage of the first or sea constituent melt A. The lower independent portion of the inlet hole 28 produces a primary composite melt stream consisting of the sea melt stream. The melt subsequently flows from a combining chamber 25 of a circumferential form, through the circumferential space 26, and the island melt stream, which flows into the sea melt through the inlet nozzle 27.

2.2.1.4. Multi-layer spinning. Liquids can be separated into different layers using static mixtures in multi-layer-type spinning [30]. Kanebo, Kuraray and Toray have done experiments with this process. Kuraray was the first to come up with a successful version. A conjugate fibre is spun from polyester and nylon 6, which is then converted into 0.2- to 0.3-denier filaments.

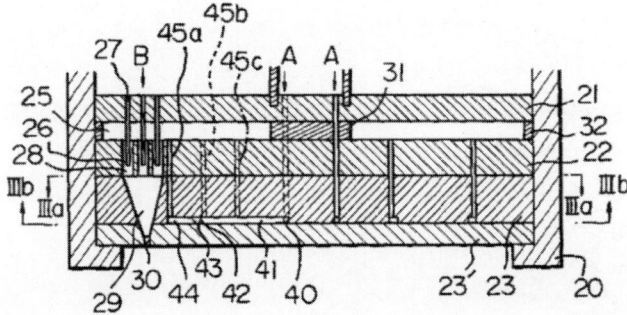

Figure 28. Improved spinneret design from Moriki and Ogasawara (after US Pat. 4445833).

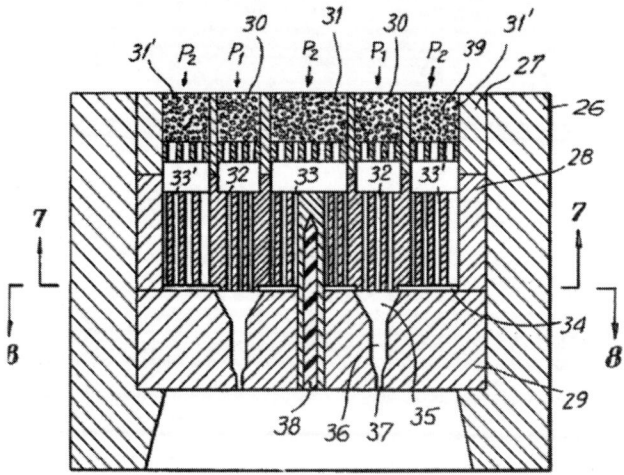

Figure 29. Spinning apparatus for multi-segment fibre (after US Pat. 4460649).

A US patent [29] discusses multi-layer type spinning and reports the development of a composite fibre consisting of two components, a polyamide and polyester. The two components of the multi-segment fibre can be subsequently separated into a plurality of microfibres by chemical and physical treatments. The fibre developed by this method has low loss of weight in a chemical treatment and can easily be separated into outer and inner components. After separation, the composite fibre gives excellent performance.

The spinning apparatus, as shown in Figure 29, comprises a pack body 26, a cup 27 for molding sand and 28 as a guide plate. The spinneret plate is shown as 29. The inner component material P_1 passes through path 30 into outlet aperture 32 in metered quantity and then enters opening 35 of guide hole 36. An important difference occurs for the second and outer component. The outer component material P_2 passes through 31 and 31′ into outlet apertures 33 and 33′ and subsequently through 34 and then enters 35 of the spinneret plate. The resulting composite melt issues from orifice 37.

Both the direct and indirect ways are being practiced today. The first one is a more economical method because of a much lower investment cost for such yarns. The latter one allows the production of very specialised yarns with interesting cross sections. A comparative study of the spinning procedures (Table 2), as summarised by Nakajima is given below.

2.3 Random (staple) type

These kind of fibres are produced mainly by the underlying methods:

(1) Melt blowing
(2) Spun bonding
(3) Flash spinning
(4) Polymer-blend spinning
(5) Electrospinning
(6) Other random processes

Table 2. Comparison of Microfibres from Different Spinning Procedures (after Nakajima [149]).

Details	Direct spinning	Islands-in-sea type	Multi-layer type
Limit of the finest denier (after splitting)	>0.1 d	˜ 0.0001 d Non-circular cross sections possible	0.1-d Flat cross section
Method	Spinning and drawing	Removal of component by solvent	Physico-chemical splitting
Processing	Relatively difficult. Different from conventional fibres. Fluffy.	Relatively easy. Similar to conventional fibres. Less fluffy.	Relatively easy. Similar to conventional fibres. Fluffy.
Inter-filament distance in fabric	Small	Can be controlled	Small
Hand	Hard	Soft	Hard unless chemically treated to shrink
Single-component fibre	Yes. Also possible to make multi-component fibre.	Yes (by dissolving). Also possible to make multi-component fibre.	No (Only possible by also dissolving).
Dyeability/Colour fastness	Easy dyeing. Poor colour development	Easy dyeing. Poor colour development	Poor colour-fastness.

2.3.1 Melt blowing

In the process of melt blowing, a jet of air blows (Figure 30) the extruded polymer from the spinneret that has a sharp edge into ultra-fine fibres that are collected as nonwovens. The melt viscosity is lower than that of the conventional polymers used for melt spinning, and an appropriate range of melt viscosity should be chosen to achieve the required result.

In conventional melt spinning, a polymer stream is ejected into a gas, usually air, at ambient temperature. The gas performs two functions: it cools the filament, and it exerts a drag force upon the rapidly moving filament. In melt blowing, as illustrated in figure, a high-velocity gas stream impinges upon the polymer as the polymer emerges from the spinneret. The substantial forwarding force obviates the need for the tension provided by a take-up roll.

Melt blowing has been popularly used for making microfibres. It is finding applications in an increasing number of fields, such as filtration, absorbency, hygiene and apparel. Many significant efforts have been made to better understand the technology and to improve the equipment by researchers and engineers around the world. Wadsworth and Muschelewicz [31] reported that for a polymer throughput of 0.2 g/hole/min of PP, a significantly smaller fibre diameter was observed at a high airflow rate than at a low airflow rate. Haynes [32] found a correlation between the average PP fibre diameter and the momentum of the air jet, i.e. higher air jet exit momentum per unit die span will result in smaller fibre diameter.

Yin et al. [33] showed that the PP filaments were attenuated most quickly at a distance of 5.08–7.62 cm from the die tip, after which the fibre diameters did not change significantly. Uyttendaele and Shambaugh [34] have developed a model for steady polymer melt blowing using PP with a single-hole, concentric die by following the earlier studies of conventional

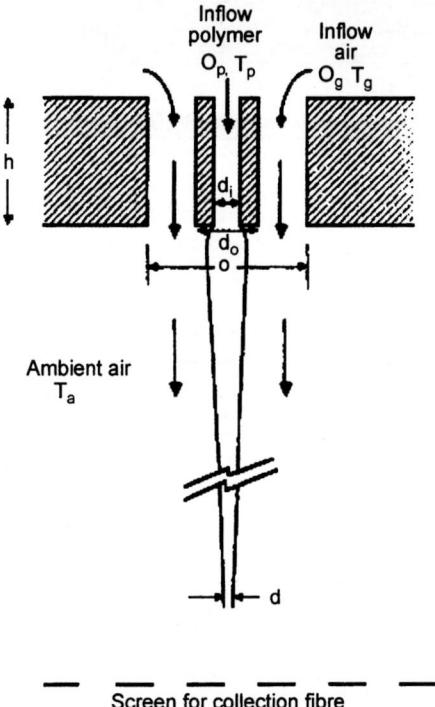

Figure 30.   Schematic diagram of melt blowing.

melt spinning. Milligan and Haynes [35] also conducted modelling research on PP melt blowing based on a single-hole die with the air impinging upon the filament at an angle of 30″ from two sides of the spinneret nozzle.

2.3.2 Spun bonding

The usual spun-bonding process consists of laying down a web of continuous filaments and subsequent bonding of fibres using mechanical, thermal or chemical means. It has to be noted that the melt blowing process also fits the general definition of a spun-bonding process. Conventional spun-bonding method is generally used to produce normal denier fibres. Spun bonding has also been used to produce microfibres. A US patent [36] describes the formation of microfibres.

The system looks like the conventional spun-bonding system. However, the filament-drawing system uses a multi-level quench (Figure 31), and uses a portion of the quench fluid as a part of the drawing. The drawing system comprises a jet with adjustable primary and secondary nozzles with a variable-width draw jet slot. The entire draw jet assembly is movable vertically for filament optimisation. The constant-flow secondary nozzle produces a very high velocity increment to the filaments by oscillating the fibres and resultant fibre denier of up to 0.5.

2.3.3 Flash spinning

In the flash spinning technique, solution of an organic polymer under high pressure and temperature far above the boiling point of the solvent is extruded into an area of substantially

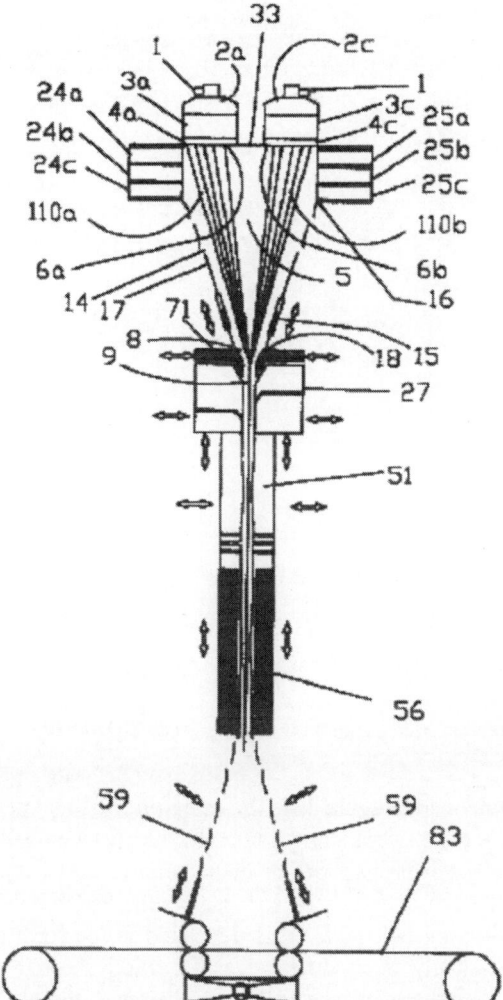

Figure 31. Schematic diagram of spun bonding (after US Pat. 6379136).

lower atmospheric pressure. As the material issues from the orifice, the solvent 'flash' evaporates and a plexifilamentary strand is formed. The strand is composed of very thin film fibril elements, which are interconnected in a three-dimensional network.

The flash spinning process requires a solvent with the following characteristics:

(1) It is a non-solvent to the polymer below the spin agent's normal boiling point.
(2) It forms a solution with the polymer only at higher pressure.
(3) It forms a desired two-phase dispersion with the polymer when the solution pressure is reduced slightly in the let-down chamber and flash vapourises when released from the let-down chamber into a zone of substantially lower pressure.

In the flash spinning technique, as the material comes out from the orifice (24), the solvent flash evaporates and a plexifilamentary strand is formed (see Figure 32). The plexifilamentary strand is composed of very thin film fibril elements that are interconnected

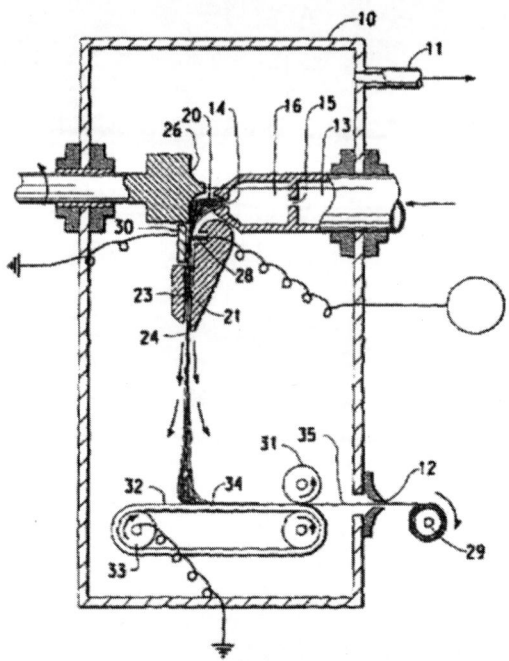

Figure 32. Schematic diagram of flash spinning (after US Pat. 6638470).

in a three-dimensional network (Figure 33). The three-dimensional network is spread into a wide web by causing it to be swept along a smooth path past a curved surface baffle plate (21) where the expanding solvent gas spreads the material.

By oscillating the deflecting baffle, the web can be directed to various areas across the width of a moving collecting belt where it is deposited in the form of swaths. The web can be further electrostatically charged to increase its width through mutual electrostatic repulsion between fibrils and attract the swath to the belt and immobilise the deposit. The

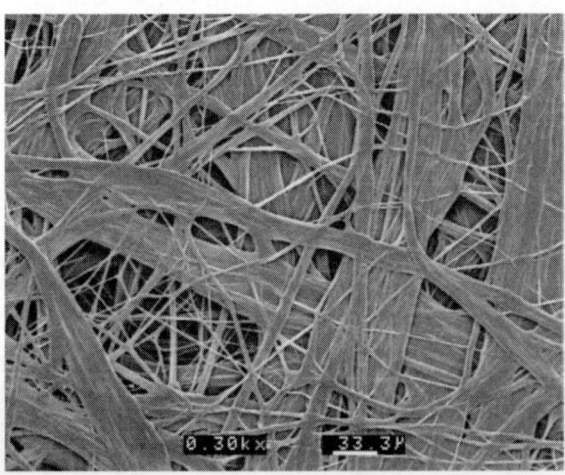

Figure 33. A micro-networked structure formed out of flash spinning technique [37].

resultant material is a fibrous sheet. Then, the sheet is subjected to either area bonding, which creates a stiff, paper-like sheet, or point bonding, followed by in-line softening that creates a drapeable, fabric-like sheet. No binders or fillers are used in the flash spinning process [38].

This technology was found by Du Pont accidentally while examining the explosion behaviour of organic solvents for safety research. As an example of the process, polyethylene is dissolved in hydrocarbon or methylene chloride, heated under pressure and jetted from a nozzle into a micro-networked fibre turned into a 'plexifilament'.

Ethylene chloride and fluorocarbon have been used as solvents, but they are now being replaced with other solvents that will not destroy the ozone layer. This technology was first aimed at producing pulp for synthetic paper, but has been converted to produce wrapping materials including wrappings for domestic use and envelopes. The spinning speed is too high to reel in the product in the form of fibre, so the product is made into a sheet.

2.3.3.1. Developments in flash spinning technique. Since its inception, the flash spinning process has undergone some developments. In a process described in Pollock and Smith and colleagues [39], the oscillating baffle in flash spinning was replaced by a rotating baffle, having specially contoured surfaces, which simultaneously spreads and oscillates the web as it is directed through an electrostatic device to apply uniform electrostatic charge on the web and promote uniform deposition of the web on a moving collecting surface. The apparatus had an annular disc target electrode that is concentric with the rotating baffle and rotates independently of said baffle. A multi-needle ion gun was positioned opposite the target plate, the needles being aimed at a portion of the target electrode to provide a corona-discharge zone. The fibrous material moving in a planar path between the target electrode and the ion gun needles was electrostatically charged before being deposited on the moving collecting surface.

The uniformity of the web from the flash spinning technique was a major concern for long. A number of requirements must be satisfied to obtain wide, fibrous, nonwoven sheets having a uniform appearance and a uniform basis weight. In general, wide nonwovens are obtained by blending and overlapping the swaths from several spinning positions. Smith [40] describes a mechanism for making fine adjustments and varying the weight distribution of the swaths deposited on the collection surface. Tests have shown that uniformity in the cross machine direction (i.e. the direction at right angles to the direction of movement of the receiving surface) is obtained when the width of the swaths at this surface is within certain limits. This, in turn, depends on the shape of the cross-machine direction basis weight profile of each swath. The width is a function of the amplitude of the oscillation imparted to the web by the baffle, the amount of electrostatic charge on the web and the distance between the baffle and the receiving surface.

Isakoff [41], describes a short diffuser device, termed a 'scoop', which 'squeezes' the gaseous stream to increase its width, thereby also increasing the width of the entrained web. This scoop was situated between the baffle and the corona-charging station, and leads to improvements in sheet uniformity by permitting shorter baffle-to-receiving surface distances for a given swath width. With increase in spinning throughputs, the larger volumes and velocities of gas produced in the flash spinning operation create an undesirable increase in turbulence. This increases the random oscillation of the web producing a nonwoven sheet having less than the desired uniformity.

2.3.4 Polymer-blend spinning

In this method, the conjugate fibre is produced by extruding and drawing a blended polymer melt of two components. The arrangement of the dispersed and non-dispersed (matrix) components is determined by the mixing ratio of the components and their melt viscosities. UCC and others found that discontinuous ultra-fine fibres could be obtained by removing the matrix component while investigating gut extrusion. Fukushima el al. successfully applied this method in the production of artificial leather. A conventional spinning facility can be easily converted for this type of polymer-blend spinning by adding a mixer-extruder. Here, the fibre fineness cannot be controlled and the fibre often breaks during spinning, although the spinning stability is strongly dependent on the combination of polymers. Since the dispersed polymer phase is drawn to yield ultra-fine fibres, no continuous-filament type of ultra-fine fibre is produced at present by polymer-blend spinning.

2.3.5 Electrospinning

The process of electrospinning uses a high-voltage source to produce an electrostatically driven jet of polymer solution that thins and elongates as it is driven towards an electrically grounded target. The fibre diameters that result from this process are on the order of nanometres to micrometres and can have large surface area to weight ratios.

A wide variety of polymers [42–44] have been electrospun into nanoscale- or microscale-diameter fibres using a simple laboratory instrument that utilises significantly less material than required for conventional melt or solution spinning. Polyethylene microfibres have been electrospun both from the melt at 200°C–220°C and from a dilute solution in paraffin at 100°C by Larrondo and Manley [45–47]. The polyethylene fibres produced had fibre diameters of several microns up to several hundred microns.

Electrospun linear low density polyethylene (LLDPE) fibres manufactured by Givens et al. [48] had diameters of 2–7 μm (Figure 34) and possessed a roughened surface morphology. This is a slightly larger diameter than observed for most electrospun fibres. During electrospinning, it was noted that there appeared to be less fibre-whipping action than observed previously for other fibres. The reduced whipping action could well be the source of the larger diameter fibres due to reduced drawing during spinning. Another possibility could be more rapid onset of crystallisation as the solution cools after leaving the Taylor cone. This would also lead to reduced draw during spinning.

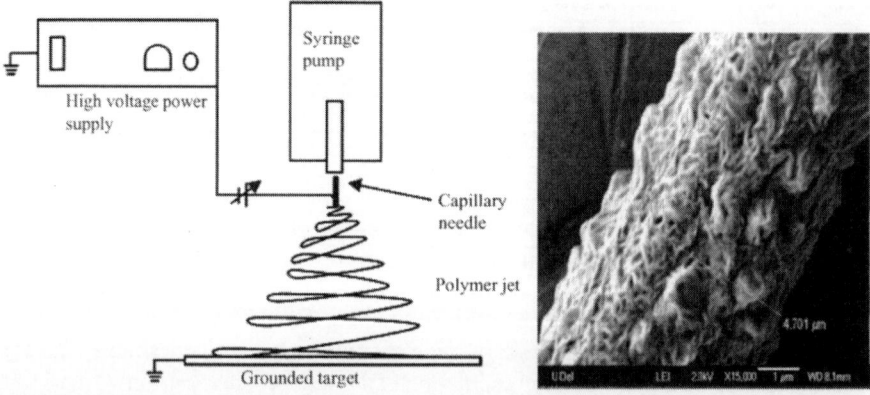

Figure 34. Electrospinning: (a) the set up, and (b) SEM of polyethylene microfibre produced (after Givens et al. [48]).

Figure 35. Scanning electron micrographs of electrospun microfibres consisting of (a) 2 μm, (b) 3 μm, (c) 4 μm, (d) 5 μm, (e) 6 μm, (f) 7 μm, (g) 8 μm and (h) 10 μm fibre diameters. The scale bar shown applies to all images and is equal to 10 μm (after Pham et al. [49]).

In a study by Pham et al. [49], poly(ϵ-caprolactone) microfibre scaffolds with average fibre diameters ranging from 2 to 10 μm (Figure 35) were individually electrospun to determine the parameters required for reproducibly fabricating scaffolds. As fibre diameter increased, the average pore size of the scaffolds, as measured by mercury porosimetry, increased (values ranging from 20 to 45 μm) while a constant porosity was observed.

To capitalise on both the larger pore sizes of the microfibre layers and the nanoscale dimensions of the nanofibre layers, layered scaffolds were fabricated by sequential electrospinning. These scaffolds consisted of alternating layers of poly-(ϵ-caprolactone) microfibres and poly(ϵ-caprolactone) nanofibres (Figure 36). The authors commented that the quantity of nanofibres and microfibres in these multi-layer structures could be balanced, and structures with tremendous potential for 3D tissue engineering applications can be produced.

2.3.6 Other random processes

Other random-type spinning processes majorly researched by Okamoto [50–55] for ultrafine fibres are:

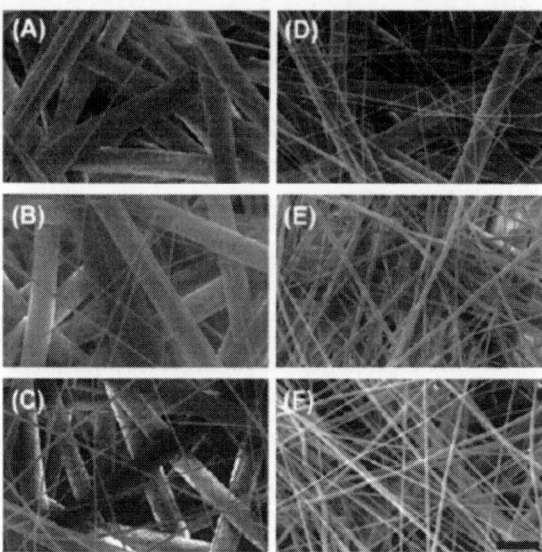

Figure 36. Scanning electron micrographs depicting the effect of different periods of nanofibre electrospinning times on the accumulation of nanofibres on top of an existing microfibre layer (5 μm fibre diameter): (a) 15 s electrospin, (b) 30 s electrospin, (c) 60 s electrospin, (d) 90 s electrospin, (e) 120 s electrospin and (f) 300 s electrospin. The average fibre diameters are 610 $\pm$ 120 nm and 5 μm for the nano- and microfibres, respectively. The scale bar shown applies to all images and is equal to 10 μm (after Pham et al. [49]).

(1) Ultra-centrifugal spinning, where ultra-fine fibre is spun like a cotton candy.
(2) Fibrillation by beating, where a fibre or film is beaten to fibrillate it.
(3) Fibrillation by turbulent flow, where a polymer in solution is coagulated in a zone of turbulent flow.
(4) Burst spinning (blown sheet method), where a blowing agent or gas is introduced into the polymer to burst it apart and form ultra-fine staple fibres.
(5) Surface-dissolving method, where the surface of PET fibres etc., is dissolved in alkaline solution and the fibres are thereby made thinner.
(6) Ultra-fine fibre formation is also affected by whisker production, by reducing weight by carbonisation, tack spinning and high-pressure water-flow techniques.

3. Microfibres from different polymers

Microfibres have become almost synonymous with polyester and nylon. Trevira Finesse, Fortrel Microspun, DuPont Micromattique and Shingosen [56] are all trade names for various polyester microfibres, whereas Supplex Microfibre, Tactel Micro and Silky Touch are some of the trade names for nylon microfibres.

Although the majority of the experiments and developments have taken place with polyester and nylon, there have been interesting developments with other fibres too.

3.1 Acrylic

Montefibre [57] has been one of the first synthetic fibre producers active in developing and producing acrylic microfibres. A new polymer and a new continuous spinning process have

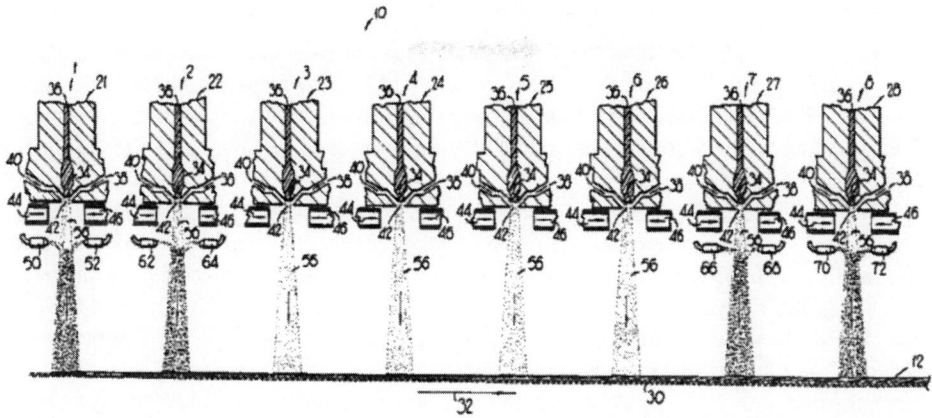

Figure 37. PP hydrophobic microfibres sandwiched between hydrophilic microfibres (after US Pat. 4753843).

been developed to produce acrylic microfibres Myoliss and Leacril Micro with filament counts of 0.9–0.8 and 0.6 dtex without losing spinning machine capacity and quality fibre consistency compared with the high filament counts production. Myoliss for the worsted system and Leacril Micro for the cotton system, both ring and open-end (OE), have an excellent fibre-to-yarn processability, allowing the production of fabrics highly innovative for soft touch, aesthetics and for wear comfort to satisfy the needs of a more sophisticated market. Montefibre is pursuing the development in this field with new types of Leacril Micro with improved fibre-to-yarn processability and with new performances on final fabrics.

3.2 Polypropylene

A US patent [58] discusses a multi-layered, absorbent, protective, nonwoven web that has one or more central layers of melt-blown polypropylene microfibres that are naturally hydrophobic.

The central layers are sandwiched (Figure 37) between one or more melt-blown surface layers on each side. The surface layers are composed of melt-blown polypropylene microfibres that have been rendered hydrophilic by the addition of a non-ionic surfactant during the formation of the surface-layer microfibres.

3.3 Cellulose

There has not been much research on cellulosic microfibres, but some interesting applications have been reported.

Fibrous absorbent webs having a low density (about 0.01–0.15 g/cm^3), and comprising of at least 50% cellulose microfibres (CMs) having a diameter of about 0.01 to about 15 μm has been reported in a US patent [59].

In a study by Toledano-Thompson et al. [60], henequen CMs were grafted in a two-step method with polyacrylic acid (PAA). Initially, the CMs were treated with an epoxide that contains a terminal double bond. The epoxide reacts at the surface with the cellulosic fibres that became functionalised with a terminal double bond. It was followed by the grafting of PAA onto the cellulosic fibres using a solution polymerisation reaction initiated with potassium persulphate. Thermogravimetrical analysis showed that the grafted CM with

PAA had a better stability than the CM treated with epoxide and cellulosic microfibres. Sorption measurements confirm that PAA-grafted CM had a water-sorption capacity three times larger than that of the CM without grafting. The increase in water sorption is attributed to the presence of PAA in the fibre rather than an improved capillarity of the cellulosic fibres.

CMs were isolated by Bhattacharya et al. [61] from bagasse in three distinct stages. Initially, bagasse was subjected to established pulping process to eliminate lignin and hemicellulose. The obtained cellulosic fibres were mechanically separated into their constituent microfibrils (MFs) by a two-stage homogenisation process and were, finally, acid hydrolysed. The dimensions of the resulting microfibres, as commented by the authors, were dependent on the hydrolysis conditions. Persistent discoloration indicated that cellulose obtained from bagasse, a sugarcane by-product, was far more resistant to hydrolysis than tunicate, bacterial or even wood celluloses. Solid-state nuclear magnetic resonance (NMR) spectroscopy of the cellulose fibres confirmed that lignin had been completely eliminated during the pulping process. The ^{13}C NMR spectra of the MFs also clearly indicated that hydrolysis and mechanical shearing resulted in significant removal of the amorphous regions that were initially present in the unhydrolysed cellulose fibres. This was a significant research reporting of getting purely crystalline cellulose in the form of microfibres.

4. Developments in microfibre manufacturing

4.1 Changing the cross section without changing the spinneret

Fibres of virtually any cross-sectional geometry are formed by melt-spinning fibre-forming polymers through specially designed spinnerets. Thus, to obtain fibres of a specific cross-sectional geometry, a corresponding spinneret orifice of specific geometric design is typically needed. An interesting exception to this has been reported in a patent. According to the patent [62], at least two polymers can be co-melt-spun through an orifice of fixed geometry so as to achieve a bicomponent fibre having a desired cross section. To change to a bicomponent fibre having a cross section that is different, at least one of following is to be changed:

(1) the differential relative viscosity between the first and second polymers,
(2) the relative proportions of the first and/or second polymers, and
(3) the cross-sectional bicomponent distribution of the first and second polymers.

A wide variety of bicomponent fibres having different cross-sectional geometries may be produced without changing the fixed-geometry orifice through which the polymers are co-melt-spun. Thus the bicomponent fibre cross sections may be 'engineered' to suit a variety of needs without necessarily shutting down the production of fibre-spinning equipment to change spinnerets.

4.2 Innovative way to make nonwovens from cellulosic microfibres

The institute TITK (Thüringisches Institut für Textil–und Kunststoff-Forschung), Rudolstadt succeeded in spinning cellulosic microfibres by specific mechanical shear deformation of cellulose/N-methylmorpholine-N-oxide solutions. The presence of viscous desolvation media consisting of solutions of fibre-forming polymers in dimethylformamide (DMF) or dimethyl acetate (DMAc) is a necessary condition. Cellulosic microfibre nonwovens can be manufactured by simple filtration from the microfibre dispersions. It is also feasible to spin the microfibre dispersions to matrix-fibril-fibres, whereby the cellulosic microfibres are positioned longitudinally to the fibre axis. These matrix-fibril-fibres can be processed to microfibre nonwovens by traditional textile methods.

4.3 Radial quenching system

With the radial-outflow quenching system [63], the loss in capacity was avoided. A suitable modification of the standard system leads to the most economic and optimal production conditions for the processing of microfibres. Also, in the microfibre range, the advantageous quality figures, in comparison with other cooling devices, are achieved with the use of radial quenching device.

4.4 A new self-suction cooling device

Akzo Nobel Faser AG, Germany [64] have developed a new process for the spinning of microfibre POY yarns. The process utilises a self-suction device to cool the filaments after melt extrusion. Details are provided of work carried out on a multiple-position pilot unit to assess its potential for spinning filaments from 0.4 to 1 dtex including the design of the cooling unit, yarn path, application of spin finish, spinneret design and the properties of the yarns obtained. It was found that the process offered excellent spinning reliability.

5. Texturing machines for microfibres

With the production of microfibres, the demand to texture such fibres quickly became a reality. But to the consternation of the texturing industry, microfilament yarn seemed to be very difficult to texture. Broken filaments and the inability to utilise the yarns on high-speed looms were the major setback. ARCT (today ICBT) of France, introduced years ago in a Textured Yarn Association of America (TYAA) meeting in Myrtle Beach yields test results of an improved machine design. The trick was to have basically a single curved-texturing zone and cooling zone without any major deflections. This concept was later picked up and further refined by nearly every other texturing machine producer. Now interlacing the finished textured micro yarn is an accepted standard, allowing much improved weaving, warping and knitting speeds.

Microfibre is more heat sensitive, and hence, requires precautions like straight and short yarn path in texturing zone, specially designed low-friction ceramic guide surfaces and specially designed polyurethane friction disks for an even-twist insertion. The fibre should be processed at 20%–30% lower texturing speed, no second heater treatment is necessary because of the fibre's low crimp modulus and micro-coning oil to cater to the special needs of the wicking action of microfibre.

A typical example for machine set-up for microfibre made from POY is given as under:

Draw ratio	: 1.59
First heater temperature	: 180°C
Friction disks	: One ceramic entry disk, one ceramic exit disk
DY ratio	: 1.95

6. Mechanical processing of microfibres

6.1 Introduction

Microfibre properties are influenced in many interesting ways as dpf (denier per filament) is reduced. These changes of properties may affect both processing conditions and potential end uses. A reduction in dpf has an immediate impact on fibre flexibility, which in turn increases difficulties in processing as there are more chances of nep formation and fibre breakage at each stage where fibres are manipulated and requires a reduction in number of steps in processing. This increase in flexibility due to the reduction of dpf is related to the reduction of bending rigidity. The reduced bending rigidity of microfibres results in fibre damage in carding process. There are several advantages of producing fabrics from microdenier fabrics as far as fabric properties are concerned, but due to the extra fineness of fibres, critical problems are confronted during processing.

End products incorporating microfibres must satisfy functional requirements like wind-proofness, water repellents apart from fashion attributes like drape, soft handle and appearance. Accordingly, microfibre yarns have sound applications in sportswear, women's and men's fashion outer layer and household fabrics. The great majority of these end products are made from continuous filaments although microfibres are also available as cut staple fibre with lengths between 30 and 40 mm.

Before going to the details of the processing of microfibres, the properties of the microfibres that influence the spinning and the subsequent processes require a thorough discussion.

6.2 Properties of microfibres as a function of diameter

The immediate dimensional change that can be foreseen in a microfibre is a reduction in diameter. Due to the reduction in diameter, the following properties get significantly affected:

(1) Flexural rigidity,
(2) Tensile strength,
(3) Surface friction.

The diameter–fineness relationship, assuming circular cross section, would be as follows:

$$\text{Diameter } d = \frac{11.8929 \times 10^4 \times \sqrt{\text{denier}}}{\text{Fibre density}}.$$

Figure 38(a) clearly shows to what extent the properties get affected when the diameter is plotted as a function of denier, ranging from micro to normal level; Figure 38(b) gives the relation between the fibre fineness and the moment of resistance; and Figure 38(c) gives the relation between the fibre fineness and the moment of inertia.

The moment of resistance varies to cubic power and moment of inertia to fourth power of the fibre diameter. All these properties are extremely important since flexural rigidity relates to easiness with which deformation may lead to nep generation or incidence of lapping. As a fibre's resistance to bending is proportional to fourth power of the fibre radius, with a decrease in radius by a factor of 2, the bending resistance decreases by a factor of 16. The development of microfibres has thus enabled the development of fabrics with desired flexibility and softness [67]. However, it is also imminent that a small reduction in diameter through changing denier will cause a fine fibre to get deformed easily, leading to nep generation and problems in subsequent processing.

Table 3 gives important physical properties of polyester and acrylic microdenier fibres.

It can be noticed that tenacities of microfibres are lower than normal-denier fibres. Also, it has to be taken into account that the increasing fineness of fibre count is associated with the following:

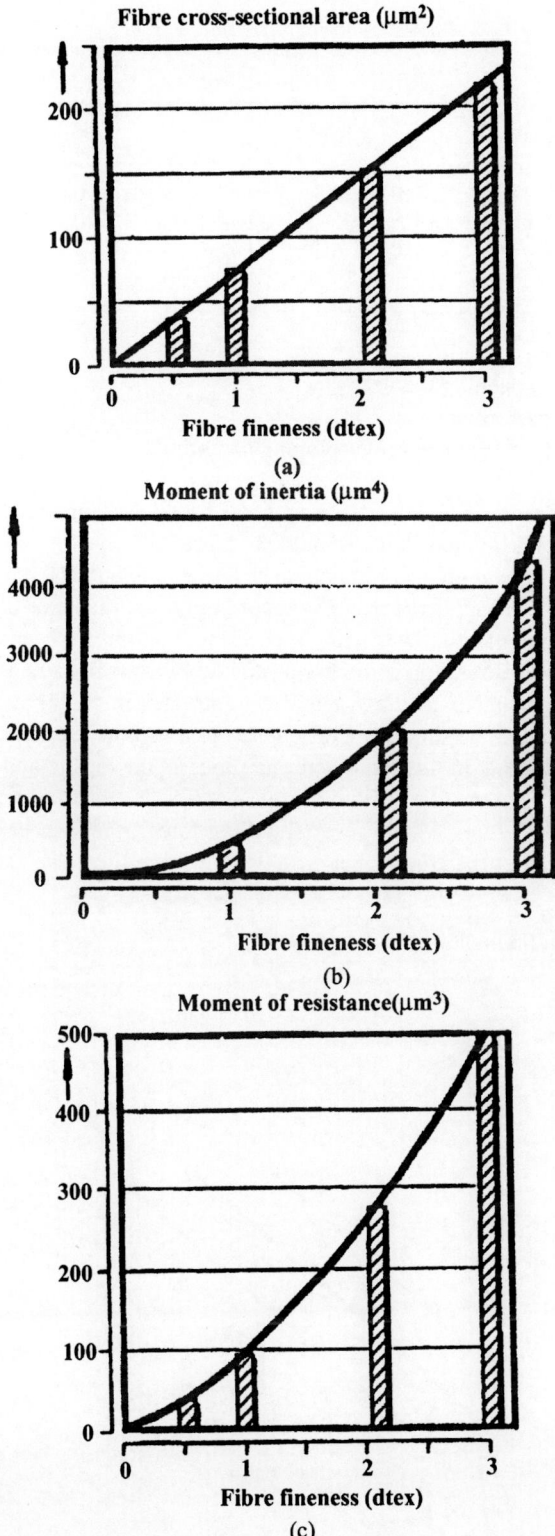

Figure 38. (a) Relation between fibre fineness and the moment of resistance; (b) Relation between fineness and the moment of inertia; (c) Relation between fibre fineness and the moment of resistance.

Table 3. Physical properties of polyester and acrylic micro denier fibres.

Type of fibre	Polyester			Acrylic		
Denier	1.3	0.8	0.5	0.5	0.8	1.3
Length (mm)	38	32	32	32	40	40
Tenacity (g/d)	6.32	5.53	4.92	2.8	3.8	3.6
Elongation (%)	19.5	19.5	24.9	20–30	30	32

(1) Greater efforts in opening the fibre stock,
(2) Lower carding performance,
(3) Higher sensitivity to unfavourable spinning conditions.

The mechanical stresses and the deformation of fibres in the manufacturing process as observed by Ferdinand Leifield [66] is shown in Figure 39.

Let's consider a fibre of diameter D, length L, tensile strength and modulus of elasticity E and the magnitude of applied force K. Accordingly, the diameter of the fibre is then the greatest determining factor. During the processing, the fibres are pressed, compressed, rubbed, drawn and bent. Pressing, compressing and rubbing are not amenable for numerical estimation. There are higher loadings that create stresses in the fibres, and this can be numerically estimated if one simply understands a circular-fibre cross section at hand. One can go from the point that in the manufacturing process, the tensile and bending strength make decisive stresses in the fibres. With the tension, it is described how a force influences the cross section. Therefore, in case of equal pulling forces, the smaller cross sections are subject to higher stresses than the higher cross sections. For the given example,

Pulling tension $= \sigma t = K/F$
Pulling deformation (elongation) $= \Delta \ell / \ell = K/EF$
Area $F = \pi \times (D^2/4)$.

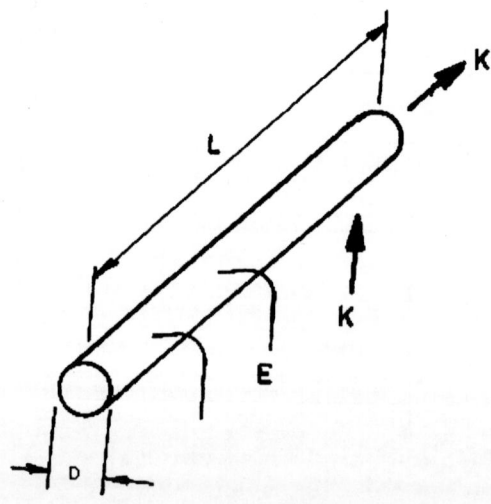

Figure 39. Mechanical stresses and the deformation of fibres.

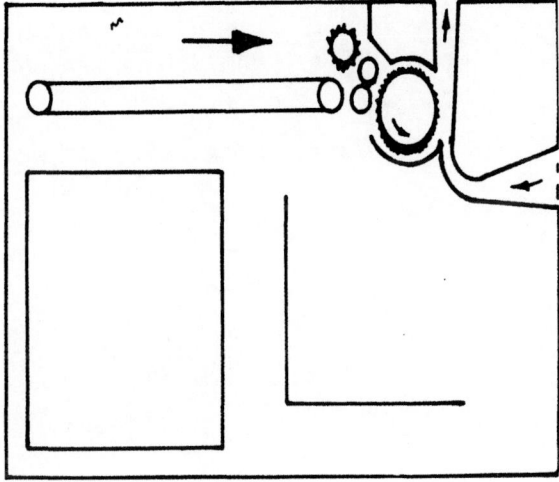

Figure 40. TUFTOMAT TFV1.

The pulling tension and the pulling deformation are dependent on the fibre cross section F, therefore, of the square of the fibre diameter. As per this, pulling tension and pulling deformation are inversely proportional to D^2.

Formula for bending

Bending stress $\sigma_s = M/W$
Bending deformation $f_{max} = L^3 K/48EJ$
Moment of resistance $W = D^3/10$
Moment of inertia $J = D^4/20$

In the case of bending for the stresses due to the force, the bending stresses and the moment of resistance W are responsible. For the bending deformation, the moment of inertia J has to be estimated.

The importance of the diameter and, hence, the microfibre can be understood from the above formulae where the variation due to diameter multiplies several times.

Ferdinand Leifield also reported in the study, *A comparative study between two fibres with varying deniers of 0.5 and 1.7 decitex*. The comparison shows how a 0.5-decitex fibre reacts in a significantly different manner than a 1.7-decitex fibre in connection with the stresses and deformations.

Table 4 shows the different stresses possible while treatment is given in the header row while the last row elucidates the possible results from such stresses. The factor shows the ratio of sensitivity of the 1.7-decitex fibre to the 0.5-decitex fibre. From the result, one

Table 4.   Stress comparison for 0.5 and 1.7 decitex fibres.

	Pulling Stress	Pulling deformation	Bending Stress	Bending deformation
Factor	3	3	5	10
Effect	Fibre damage	Fibre damage	Fibre damage	Neps

Figure 41. Opener cleaner.

can see that 0.5-decitex fibre with reference to the fibre damage due to tensile stresses and deformation reacts three times more sensitively (lesser values). As the bending deformation leads to nep formation, the fine fibre reacts 10 times more sensitively with reference to nep formation than 1.7-decitex fibres.

6.3 Spinning of microfibres

In recent years, fine-denier fibres have been developed by synthetic-fibre manufacturers for use in apparel and higher quality consumer products. However, when such fine fibres are subjected through a series of machines employed in spinning, they experience various types of stresses and strains of different magnitude. Unfortunately, the properties of microfibres that make them attractive for fabrics are also the same properties that lead to difficulties in mechanical processing as observed by Hwang and Wand [67].

A study on spinning microfibre yarns on the Murata Jet Spinning (MJS) system by Murata Machinery Limited reports that carding becomes more difficult in the manufacturing process for fine-denier yarn, and drafting becomes more difficult in ring spinning process. Also, as the yarn spins around the traveller, the tension increases leading to more end breaks if spindle speed is increased. The fibres are easily damaged, and this needs a reduction in spindle speeds and hence paves way for lower productivity. Therefore, at every stage of processing microfibres, care must be taken to maintain quality of yarns so as to produce value-added fabrics.

6.3.1 Blow room

The large number of fibres in a given cross section calls, due to reduced diameter, for intensive tuft opening. Consequently, the blow-room line must be equipped with two opening points. There are new series of openers and cleaners like the one shown in Figures 40 and 41, namely TUFTOMAT TFV1 that treats the fibres very gently, and a gentle opener cleaner. These openers are placed in the first position, where it is plucked out from the clamped feed by a full pin or needle roller.

The follower rollers should preferably be fitted with saw-tooth wire. Being sensitive fibres, the microfibres must be treated gently. Short machines, short pipe connections, short air transport and less number of machines rightly selected show the way for the solution to the problem.

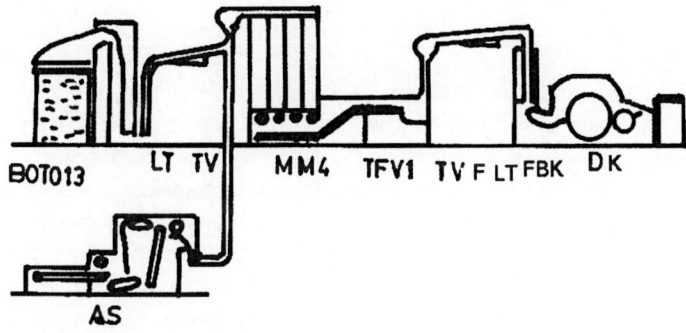

Figure 42. Blow-room line for processing pure microfibres.

Since a self-controlled stroke is necessary for a good opening, the components that pull out the fibres from the fibre tuft must be so shaped that the load on fibre is minimum. The ideal components may be pins and needles on the pin roller and needle on the licker-in. There is no other gentler component as the needles penetrate with practically no resistance in the lap because of their fine points. The round needle obstructs the sharp cutting on the edges and/or bending of the fibre. Properly selected and properly processed saw tooth are harmless, especially when they only convey or take over fibres in an open condition.

From Figure 42, it is seen that a pure microfibre blow-room installation is quite short and comprises few machines. In a typical line for processing microfibres, a BLENDOMAT BDTO13, is used for opening. Bales are fed, processed and mixed in a multimixer (blender) MM4. The waste-feeder allows the feeding of soft waste. Through the multiple blender MM4, the differences are levelled in which individual flocks from different bales on the production belt under the mixer can be brought together. After the multiple mixer, the flock feed formed on the horizontal held under the multiple mixer is lead to single roller opener TUFTOMAT TFV1 through the inclined belt. This is the most gentle and economical transfer from one machine to the following machine.

Figure 43 above shows the blow-room line for processing cotton with microfibres. The microfibres and cotton processed in rotation with a BLENDOMAT BDT019. The cotton is fed to the four roller cleaner CLEANOMAT CVT4 over multiple mixer MM4. The new cleaner CVT4 replaces a complete cleaning line of conventional design with a high cleaning efficiency. The material then reaches the dusting machine DUSTEX DX in the pneumatic weighing feeder PWSE. The microfibres are transported after the BLENDOMAT BDT 019 directly to the weighing bale opener BOVA. After mixing, it is sandwiched, processed and opened. The flocks are sent to multiple mixer MM6. As described in first installation, the flock feed leads to single roller opener TFV1 over an inclined belt and over the air distributor LT and flock feed EXACTAFEED FBK the mixture is sent to card EXACTACARD DK760.

Figure 43. Blow-room line for processing cotton with microfibres.

6.3.2 Carding

Over the last 30 years, numerous developments have taken place with the cotton card. The production rate has risen by a factor of 5 with the main rotating components running at significantly higher speeds. Triple taker-in rollers and modified feed systems are in use; additional carding segments are fitted for more effective fibre opening; and improved wire clothing profiles have been developed for a better carding action. Advances in electronics have provided much improved monitoring and process control. Most modern cards (Trutzschler DK 760, Rieter C4, etc) with presently available facilities are capable of handling microdenier fibres. In the 1980s, it was possible to produce 30–40 kg/h. The success can be shared by the fibre and machinery manufacturers. The new and accurate machines, which have the modern metallic clothing, have a remarkable share in this development.

To achieve the same throughput rate, the number of carding points in the card clothing needs to be significantly increased for microfibres. The postulations are based on the experience in the processing of cotton where the removal of neps is an essential task of the card. With polyester fibres, the formation of neps in carding must be prevented and, at the same time, fibre breakage and damage must be avoided by gentle treatment. This is certainly more difficult with very fine fibres, the ends of which tend to curl up more easily, than with coarse ones.

Hwang and colleagues in their extensive research on carding of microfibres reported a rise of number of fibres with decreasing denier per filament (dpf). A given card, thus, requires to accommodate the increase in the number of fibres. As the number of fibres increases, the openness of feed stock should be improved as the lack of openness of fibres deteriorates the web quality with an increase of nep count and fibre breakage. As suggested by the authors, to efficiently handle the increase in the number of fibres in the card, wires with high point density, high speed of card elements and proper settings become necessary. As dpf decreases, the fibre surface area per unit mass increases enormously. This, in turn, causes fibre-to-fibre and fibre-to-wire friction to increase, and it leads to difficulty in the fibre transfer from one element to another during carding. To avoid or reduce this problem, it was suggested by the authors that certain critical processing parameters be controlled such as the use of low throughput, wires with high point density and/or high speeds of elements.

Hwang and colleagues determined the cardability of polyester microfibre through the optimisation of fibre parameters, carding processing variables and their interactions. The cardability is judged by the fibre web quality in terms of nep counts and fibre breakage. Tables 5 and 6 give experimental data used, namely parameters of flat top card and levels of fibres and carding parameters of the experimental design. A polyester fibre of 0.9 dpf and a metre wide flat top card was used. An experimental design was structured to reveal the influence of fibre length, processing variables and their interactions on the cardability of microfibres. The levels of these independent parameters represented $2 \times 2 \times 3 \times 4$ full factorial design. Nep counting was done using manual template method. For each experimental run,

Table 5.   Parameters of flat top cards.

Elements	Speed (rpm)	Diameter (mm)	Wire point/in^2
Feed roller	0 – 6	100	98
Licker in	510, 930	254	40
Cylinder	190, 338	1270	684
Doffer	0–20	653	378

Table 6.   Levels of fibres and carding parameters of the experimental design.

Parameters	Units	Levels
Fibre length	mm	31.8, 38.1
Feed roller speed	m/min	0.63, 0.94, 1.26.
Draft		12.7, 19, 25.3, 31.7
Cylinder to doffer	mm	0.15, 0.30

an average of the nep count was calculated from sum of left, middle and side sections. The average of nep count was expressed in neps/g. For estimation of fibre breakage and short-fibre content, advanced fibre information system (AFIS) was used. The investigation of fibre-loading distribution in machine and cross-card direction, an infrared-based device was used. Analyses of variance (ANOVAs) were performed to test the main effects and their first-order interactions. Multiple regression analysis was performed to smooth out the data and obtain predicted equations of nep count and fibre breakage in terms of independent parameters and their interactions. Residual analysis was performed to examine the fitness of the regression model of the data.

The tendency of nep formation, which is a major problem with carding, was thoroughly investigated for microfibres by Hwang and colleagues. Figure 44 reveals the influence of fibre length on neps for different setting of cylinder to doffer. The higher the length of fibres, the more chance there is of nep formation due to lower bending rigidity. Figure 45 indicates the influence of feed roller speed on nep generation at different settings. As cylinder to doffer setting gets wider, the nep count increases. Figure 46 gives the effect of draft on neps. The nep count becomes less as draft gets higher. This is also true at different fibre lengths. One of the most discerning findings in the above research was 'that a decrease in fibre length has greater influence on the reduction of nep formation than that of the throughput'.

Effect of fibre length, feed roller speed, draft and settings on nep content for polyester microfibres

Effect of fibre length, feed roller speed, draft and settings on short fibre content for polyester microfibres

From Figure 47, one can understand that longer fibres cause an increase of short fibre content (SFC) % as longer fibres have more points of contact with fibres and wires than

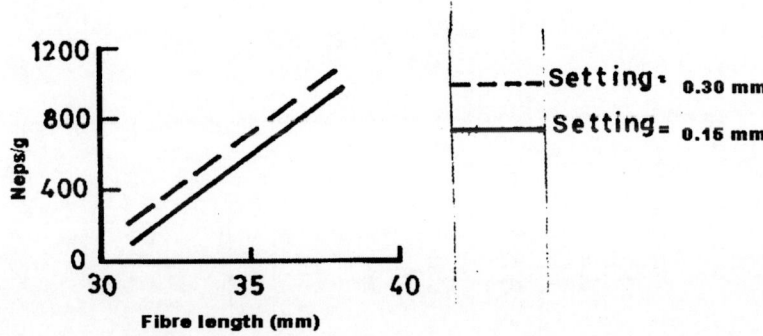

Figure 44.   Effect of fibre length and settings on nep content (after Hwang and Wand [67]).

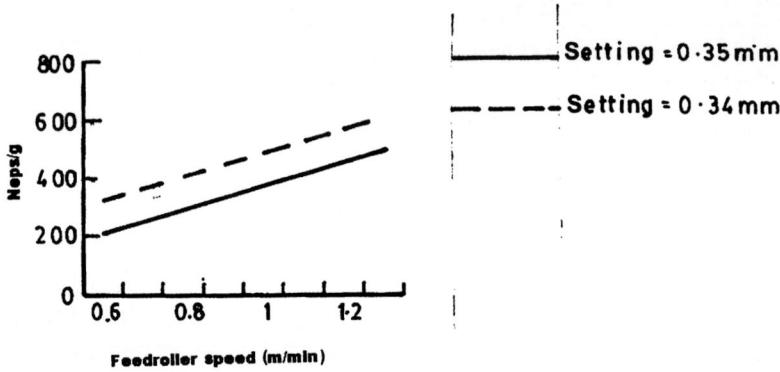

Figure 45. Effect of feed roller speed and settings on nep content (after Hwang and Wand [67]).

short fibres. This causes long fibres to entangle and break. It was also observed that SFC% increases with an increase in feed roller speed. The influence of cylinder to doffer setting on SFC% shows that as the setting gets wider, SFC% increases. The SFC% shows a declining trend as draft increases and is consistently true for different fibre lengths as observed in Figure 48.

Nep formation and fibre breakage were influenced by all main independent parameters and their interactions. The increase of draft (increase of doffer speed) led to a reduction of nep and fibre breakage, whereas the increase of throughput led to an increase of nep count and fibre breakage. Increasing doffer speed reduced recycled fibres on cylinder and contributed to less nep and fibre damage after carding. However, the increase of throughput results in the reduction of openness of fibres and an increase of nep formation and fibre breakage. The increase of cylinder to doffer settings resulted in more nep formation and fibre breakage due to the increase of recycled fibres on the cylinders. Longer fibres had higher fibre breakage on neps than shorter fibres. The shorter fibres possessed higher bending rigidity leading to less fibre breakage and neps.

It was found that there is a strong correlation between fibre breakage and nep formation. The study of nep localisation in terms of fibre load uniformity across card showed that less total neps could be produced if the fibre load uniformity in cross-machine direction is high.

Schenek and Schwippl [68] reported that card parameters are key elements for boosting performance in the spinning process with minimum fibre loading and good carding quality. To keep forces as small as possible, cylinder clothing, featuring 640 teeth/in^2 and a 30°

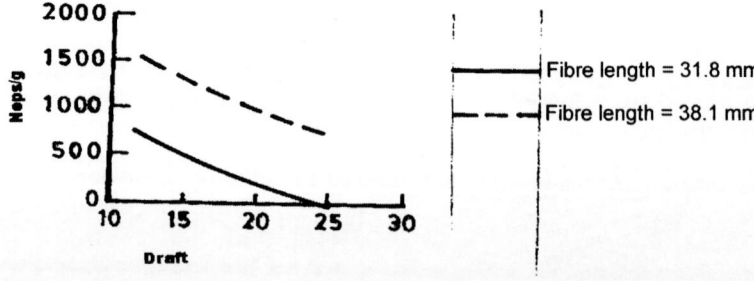

Figure 46. Effect of draft and settings on nep content (after Hwang and Wand [67]).

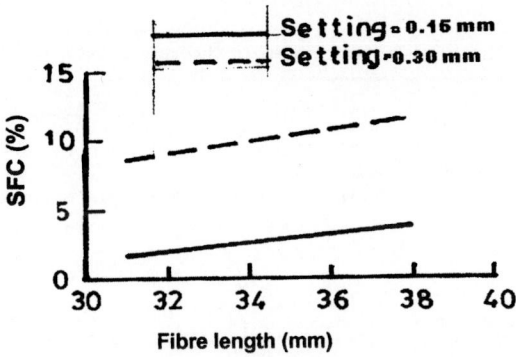

Figure 47. Effect of fibre length and settings on SFC% (after Hwang and Wand [67]).

front angle, was initially used. It became apparent that the transfer of fibre from cylinder to doffer was not ideal and that the microfibres were accumulating between the teeth of clothing. Increasing the number of teeth to 720 resulted in good running properties. The influence of fibre spinning process was also clearly demonstrated visually with gentle fibre preparation at a carding output of 30 kg/h.

A study on carding of microfibres by Leifield [69] revealed that cards with needles on the licker-in can be used for opening microfibres. However, the conventional saw-tooth licker-in can also be used for processing microfibres. Furthermore, the shape of saw tooth plays an important role and function. The experience showed that a properly dimensioned and properly processed saw tooth can go a long way with the fibre. With the cleaners as well as with the cards, the stresses in the fibres can be influenced through the setting. In openers and cards, the machine and settings were selected as used in the case of the finest cotton. For fine cotton, cylinders of 865 teeth/in^2 were selected. For processing microfibres, either the above or, preferably, cylinders with 1080 teeth/in^2 can be used. The wire breast angle 25° may be used for fibres with fewer number of neps. The flat wire density would be 450–500 points per square inch (PPSI). As far as the card cylinder speed is concerned, it must be assumed that, because of the lower centrifugal force of the microfibres, the speed should be somewhat higher to prevent the fibres from lodging at the base of the clothing.

The new EXACTACARD DK-760 allows exact and closer settings through the use of more modern aluminium extrusion press profiles for working components and coverings around the licker-in and drums. This helps in the processing of microfibres and gives safe working conditions in the production of, for example 30 kg/h. Also, at 50 kg/h, it still gives good operating characteristics. Microfibres of varying fineness (in dtex), which included

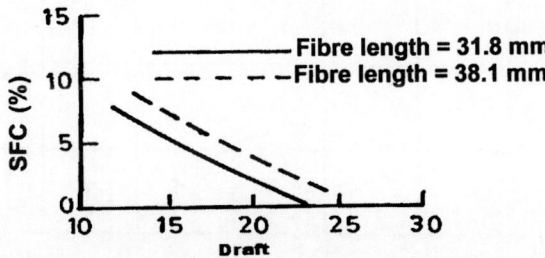

Figure 48. Effect of draft and fibre length on SFC% (after Hwang and Wand [67]).

Table 7.   Carding results of different microfibres card sliver values.

Type of fibre	Fineness (d tex)	Length (mm)	Card production (kg/h)	(Neps/g)	Average fibre length (before-after treatment)	Weight CV%	Uster CV%
Acryl Bayer Darlon	0.6	32	35	4	28–27.9	1.2	3.9
Polyester Diolen 44	0.8	40	30	1	34–32	1.5	3.4
Polyester NN	0.7	38	35	3	32–29	1.5	3.8
Polyester Skyron	0.8	38	40	1	33–31	1.3	3.4
Polyester Montefibre	0.85	38	35	3	29–28	1.5	2.7

polyester fibres of 0.7, 0.8 and 0.85 dtex and fibre lengths of 32, 38 and 40 mm were used for various experiments on the carding of microfibres for which the production levels varied from 30 to 40 kg/h. Assessments of the card sliver quality was done by estimating the neps/g, fibre damage and the average fibre length before and after carding, which are shown in Table 7 along with the weight CV% and USTER CV% for each selected card production.

From Figure 49, one can realise that from 24 to 44 kg/h, the nep count per gram falls from 7 to 3. On increasing the production to 53 kg/h, the working conditions and the other results still remain hardly influenced, but the nep count increases to 10 neps/g. This can be related to loading on the cylinder and on the previously mentioned narrow space in the intermediate of the metallic clothing. The results from the experiments carried out jointly by Bayer, Schlafhorst and Trutzschler, and reported by Ferdinand Leifield, as well as the performance of cards (Exacta card DK760) improved with the increase in production up to a certain level and then deteriorates for 0.6-denier and 32-mm Bayer Acryl fibre Dralon. At a production rate of 24 kg/h, the nep count of card sliver was 7 per gram, which became 3 per gram at 44 kg/h production. On increasing the production to 53 kg/h, the nep count increased to 10 per gram; however, it still remained in good working condition.

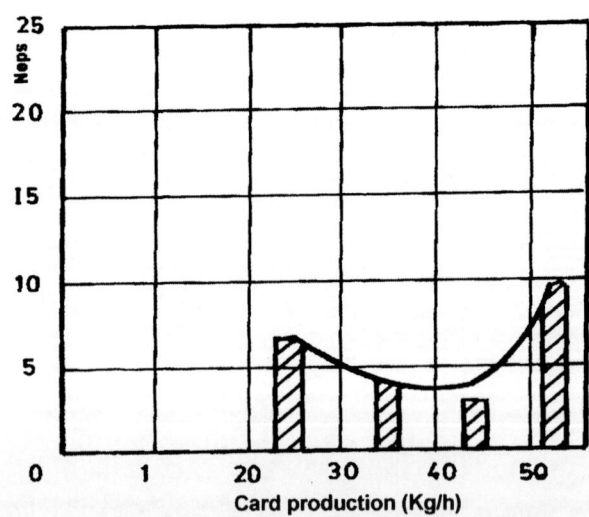

Figure 49. Neps versus card production (after Leifield [69]).

Ernst and Yarns [70] revealed that carding performance must match the fibre fineness. Throughput rates on the C4 card range from 25 to 30 kg/h depending on the fibre type and using cylinder clothing with 600–700 points/in^2. With polyester fibres, production can be increased to 40 kg/h and with acrylic up to 50 kg/h using cylinder clothing with 1000 points/in^2. On account of large number of fibres in the card sliver, it is advisable not to exceed sliver weights of 4.5 decitex.

Dr. GertBock [78] in a test on a Rieter C4 card demonstrated that using card cylinder clothing, with 860 points/in^2 and a breast angle of 25°, the number of neps in the card sliver was distinctly lower than with 640 points/in^2 and a breast angle of 9°. On the other hand, the fibre breakage increased with higher number of carding points. It was found that best results were achieved with a production rate of 45 kg/h keeping the number of carding points and the breast angle between the above two sets of values studied. Card clothing meeting the specification of 720 points/in^2 and with a breast angle of 20° was investigated on a Trutzschler DK 715 card. At a production rate between 20 and 50 kg/h, a decline in the mean fibre length about 1 mm was recorded in the card sliver but the difference no longer occurred in fibres sampled from the rotor of the spinning head. A drop in the cohesion length of the card sliver was significant, caused by the fibre crimp being drawn out at higher rates of throughput. Since no significant difference in the yarn property was revealed in the results of spinning the card sliver, rates of card throughput between 40 and 45 kg/h may be regarded as fully practicable. As far as the card cylinder speed is concerned, it must be assumed that because of the lower centrifugal force of the microfibres, the speed should be somewhat higher to prevent the fibres, lodging at the base of the clothing. In this case, the number of carding points adopted on the carding cylinder should not be excessive if fibre damage is to be kept within reasonable limits.

In case of fine thick and thin places, the C60 card system displays the same yarn values as the C51 with an 80% increase in output. At the same card output of 35 kg/h, the imperfection index (IPI) values on the C60 system are approximately 20% lower. The carding forces are, therefore, directly related to carding quality, especially when using microfibres. When processing microfibres, carding output has been limited to a maximum of approximately 35 kg/h. The new C60 carding system allows increased production. Output of up to 63 kg/h is possible, and this considerably increases the attraction of processing microfibres and sets new standards in card output, says a report from Rieter LINK 48 [71].

Murata Machinery Ltd. [72] in their study with microfibres report that carding becomes more difficult as the denier of polyester becomes finer, because nep generation during carding increases. An extremely effective means of dealing with this problem is to reduce the fibre length. By making polyester denier finer and using the MJS system, the following benefits are obtained:

(1) 20% increase in spinning speed,
(2) Increased yarn strength,
(3) Improved yarn evenness,
(4) Softer fabric hand.

6.3.3 Draw frame

For processing microfibres on modern draw frame, it may be necessary to reduce the delivery speed to 400 m/min to reduce the incidence of roller lapping, which is likely to be more because of the larger area of contact with the rollers and its low bending rigidity.

Table 8.   Details of spun yarn.

S. no.	Process parameter	Specification
1	Yarn count (tex)	13.1
2	Break draft	1.1 to 1.8
3	Roller gauge (mm)	44/56 and 44/62
4	Front top roller pressures (kgf)	14 and 10
5	Total draft	27
6	Traveller	6/0
7	Twist multiplier	3.53
8	Spinning speed	12000 rpm

Higher top roller load may be used (around 20%–25%). More frequent grinding or buffing is required. Whenever the draw frames are stopped, the pressure on the top rollers should be in released condition. Prolonged holding of fibres under pressure may cause thermal damage. The draw frames SB 851 and RSB851 can be used. Delivery speeds between 400 and 600 m/min are possible depending on the fibre type.

To establish the relationship between yarn properties and spinning conditions on the break draft, a 13.1-tex yarn was spun on a laboratory spinning frame with SKF drafting system and with parameters shown in Table 8.

Korkmaz and Behery [73] describe the interaction of specific fibre fineness values and the drawing machine in their study on drafting dynamics of fine-denier yarns.

It is important to understand the role of fibre properties in drafting process. Finer-denier fibres may have different drafting behaviour, such as drafting force, required roller settings, draft distribution and velocity change zone, than that of average-denier fibres. Fineness changes drafting behaviour by introducing fibre clusters and fibre contact points. The acceleration process depends on the interaction of static and dynamic friction forces at the fibre contact points. Therefore, the fibre movement in the drafting zone is not a continuous process, i.e. it consists of local acceleration and slowing down of segment of fibres.

Different methods have been used to observe fibre-drafting behaviour. Taylor [74] used radio-activated wool fibres to determine the proportion of fibres that accelerate to the front roller speed at any given time. Over a large part of the drafting zone, floating fibres achieve speeds greater than back roller speed for a limited time, and they have only two speeds during the transfer through their drafting zones. High-speed photography used by Taylor confirmed that floating fibres achieve speed between those of back and front rollers. McVittie and Barr [75] observed the movement of coloured fibres in an apron-drafting system with a microscope. It was found that the motion of the floating fibres was determined by frictional contact points with neighbouring fibres and the coefficient of friction between materials varied with speed. In recent years, Laser Doppler Anemometer has been used to measure fibre speed on different drafting systems with more accuracy.

Polyester fibres were used in the experiment with different deniers – 0.8, 1 and 1.2 – all with the same length of 38 mm. The fibres were in card sliver form and the sliver weight was 5 kilotex. The experiment was run at the main drafting zone of a two zone-drafting systems with three top rollers and three bottom rollers. Table 9 shows the experimental design that consisted of three different deniers of fibres, three roller settings with 6 draft combinations (3 total draft ratios with 2 break draft ratios).

Table 9. Experimental design with varying fibre fineness and roller settings.

S. no.	Property	Specification
1	Fibre fineness	0.8, 1, 1.2 denier
2	Roller Settings	43.7, 45, 47 mm

At the break-drafting zone, the roller setting was 46 mm and the break-drafting ratios were 1.47 and 1.76 with a constant 1.80 m/min incoming speed. The three total drafts used in this study were 6.07, 6.8 and 7.73. An Olympus encore 2000 model high-speed camera with a motion analyser was used to record fibre movement along the drafting zone. The data collected were tested by ANOVA, and the mean separation was performed by the least significance difference at $P = 0.05$ if the F test was significant at the same level. The results are tabulated in Table 10.

The analysis of results from Table 10 reveal that the average fibre speed in the main drafting zone depends significantly on fibre fineness that accounted for 22% of total variation in fibre speed. The average fibre speed increased with increasing fibre fineness. The microfibre had a speed of 9.95 m/min, whereas 1.2-denier fibres had 7.6 m/min as an average speed. The draft combination had a significant influence on fibre speed accounting for 18% of total variation in average speed. The interaction of setting and draft ratio was a significant factor and accounted for 7% of the total variation. Fibre fineness and draft ratio significantly affected the CV percentage of speed, accounting for 10% and 9% of the total variation, respectively.

The statistical analysis revealed that fineness has a significant effect on average fibre speed and introduces the clustering effect and different numbers of fibre contact points. The number of fibres in the cross section of microfibre cross sliver was 50% more than that of a 1.2-denier card sliver. Hence the crowded structure of the microfibre sliver increased the number of contact points. The more direct effect of fibre fineness on the cluster structure is bending rigidity that depends on the shape factor of the fibres and is proportional to the fourth power of diameter for round fibres. Microfibres have a lower bending rigidity that allows them to bend or wrap more easily than coarser fibres. Thus, the clusters formed by

Table 10. Sources of variation in the analysis of variance for the effect of fibre fineness, draft ratio and setting on the speed and CV% at the main drafting zone.

	Speed (m/min)		CV% of speed	
	p	Contribution (%)	P	Contribution (%)
Fineness (F)	< 0.001	22	< 0.001	10
Draft Ratio (DR)	< 0.001	18	< 0.001	9
$F \times$ DR	< 0.001	16	< 0.001	13
Setting (S)	NS	< 1	0.049	2
$F \times S$	NS	< 1	NS	2
DR $\times S$	0.004	7	NS	2
$F \times$ DR $\times S$	NS	5	NS	9
Error	—	31	—	53
Total	—	100	—	

microfibres are more compact than the coarser fibre clusters, due to higher compression forces created by flexible fibres.

The closer packing of microfibres results in a high number of contact points. Each fibre contact point is exposed to a different magnitude of friction force; the forces determine the local fibre acceleration and the drafting force. In the experiment, microfibres have the highest average speed that decreased with increasing fibre denier. From the statistical analysis, 43.7-mm setting had the lowest average speed, whereas 47-mm setting showed the highest speed. Even though the difference between fibre fineness levels were small, the interaction of fineness with drafting conditions was highly important. It was found that microfibres had the highest speed but the lowest variation compared to other fineness levels. Therefore, machine adjustments should be carefully made especially when working with fine-denier fibres to spin high-quality yarns.

Korkmaz [76] made a thorough investigation on the effect of fine denier polyester fibre fineness on dynamic cohesion force and report that one of the most important parameters regarding the processability of synthetic fibre is the level of inter-fibre friction, which determines the drafting force or cohesive force dependent on such factors as surface condition, crimping and lubricant deposit on the fibre. Staple length and crimp are highly significant factors affecting the cohesion values. Experiments conducted on cotton fibres revealed that there was a positive correlation between 2.5% span length and variation in cohesion force that affected ends down in spinning. Fibres with smoother surfaces had greater cohesion than those with geometrically rough surfaces. Crimp drastically influences the cohesive force by changing the surface roughness and the cross-sectional shape of fibres. Studies have been carried out on cohesive force and its influence on yarn quality, and their results indicated a strong interaction between the variability in drafting force and the variability in the single-yarn strength. The inter-fibre friction affects the drafting force that has to be overcome in every drawing section of production. This force has to be controlled to reduce any drafting problems.

The study by Korkmaz [76] illustrates the effect of fibre fineness on dynamic cohesion force at different delivery speeds. Polyester fibres were used in the experiments, and all the materials had the same fibre properties apart from fibre fineness. Polyester fibres of 0.88, 1.11 and 1.33 decitex – all having the same length of 38 mm – were used in the study. The fibres were in the card sliver form with a sliver weight of 5 kilotex. A cohesion metre R-2020 was used for the dynamic cohesion testing. When the material was passed through a drawing-frame-like device, the sliver was exposed to a draft, and its resistance to drawing was electronically measured by the electronic tensiometer. The testing time was 1 min with a 50 reading/s rate and the machine setting was 90 mm long. The test was run at three different speeds of 15, 30 and 60 m/min, and the draft ratio was maintained at a constant of 1.25:1. After each test, sliver irregularity was measured, and the collected data was tested by ANOVA.

The main effect of fibre fineness accounted for 76% of total variation in dynamic cohesion force. The delivery speed by itself was not a major factor, contributing only to 5% of total variation. The fibre fineness did interact with delivery speed accounting for 13% of total variation in cohesion. There was a curvy linear relationship between fibre fineness and cohesion, and the cohesion increased as the fineness increased from 0.88 to 1.1 decitex, after which it levelled off at all delivery speeds. The cohesion force of microfibre sliver decreased linearly with increasing delivery speed, whereas for slivers of 1.11 and 1.33 decitex, cohesion force increased linearly with an increase in delivery speed from 15 to 16 m/min. These findings indicate that the sliver structure of microfibres differs from that of other two fibres. Even a small change in fibre fineness altered the drafting behaviour drastically due to cluster formation or fibre grouping, and the cluster formation depends on the number of fibres in a cross section and the bending rigidity of the fibre. Microfibres have a high number of fibres in a cross section, but lower bending rigidity values that result in more compact clustering or grouping of fibres in the sliver and higher number of fibre contact points.

6.3.4 Roving frame

Due to higher cohesive force between the fibres, the twist level used may be a little less as compared to normal-denier fibre (i.e. TM 0.85–0.9). Higher top roller pressure should be used. The top rollers need to be buffed more frequently. Flyer with highly polished surface may lead to more fly generation and, hence, a matt finish would be preferable. Fly top with more number of ribs may be more advantageous.

6.3.5 Ring frame

The sliver and drafted fibre web being very thin, care has to be taken so that proper drafting takes place. Use of softer cots may be useful as the fibres will be gripped better, thereby reducing slippages. Higher break draft or wider setting (back zone) may be required due to higher cohesiveness of roving. For fibres finer than 0.5 denier, very high spindle speed may be avoided as it is likely to damage the fibre. The traveller speed should be restricted to 30–35 m/s along with smaller ring and shorter lift. For 0.8-denier polyester fibre, the experience shows that the yarns can be spun at the same spindle speed as conventional fibres. In addition, it is possible to reduce the twist of the yarn due to the presence of higher number of fibres per cross section. The results of micro-modal yarn spun on a ring spinning system show very good properties as compared with normal viscose fibre yarn. The count strength product (CSP) and run in kilometer (RKM) are higher, and a lower count CV% as observed by Basu et al. [77].

6.3.6 Open-end rotor spinning

Microfibres are available between 0.6 and 1 decitex. Arithmetically, this should allow spinning yarns down to 8 tex (NM 120/NE 70) with 100–110 fibres in the cross section. Microfibres based on polyester, viscose/modal and polyacrylic are majorly spun while polypropylene and polyamide are less commonly used in rotor spinning machinery. The commercial availability of these fibres and the interest shown in them prompted researchers to investigate their suitability for rotor spinning as well as their processing behaviour at high rotor speeds, spinning limits and the end products made from the yarns. Apart from the above, the objective was to find appropriate finished articles for the extremely fine rotor yarns spun from microfibres. The results obtained in both yarn and end products show that the use of microfibre in either 100% or blended forms opens up an opportunity to spin finer yarns that were not able to be achieved in rotor spinning with standard fibres. Outstanding yarn quality, soft handle in the finished product and superior aesthetics can thus open up new applications for rotor yarns as reported by Hainz Ernst in the study on OE spinning of microfibres. Systematic experiments with linear densities of 1.1 and 1.6 decitex revealed increasing fibre damage in a relatively large number of fibres in the draw-frame sliver. This knowledge may be extended to microfibres [78] especially at a relatively higher rotor speed, a noticeable feature of heavy-feed sliver is a high fibre-length-variation index. If the fibres in the rotor groove are inspected, in the heavier slivers, an increasing proportion of short fibres are found. The risk of fibre damage may, therefore, be mitigated by fine slivers. With the industrial standard draft of up to 200 into rotor machine, the feed sliver would have to be very fine. For example, with tentex yarn (NM 100/NE 60) spun with a draft of 200, a sliver of 2 kilotex is required. By optimising the sliver speed and pneumatic fibre transport in the Ri-Q-Box spinning unit of the Ru14-A machine, much higher drafts are possible. A standard draft range up to 300 is available with this machine. During trials, drafts up to 400 were used.

GertBock [78] reported that, in case of OE rotor spinning, microfibres played an important role as more number of fibres are required in the cross section because of the inferior material strength utilisation. In case of conventional ring-spun yarns, 53 fibres are required in the yarn cross section, whereas OE rotor yarn needs 90 fibres. The microdenier polyester fibres fulfill the following aspects:

(1) To spin very fine OE rotor yarns, i.e. finer than 10 tex in 100% polyester and finer than 12.5 tex in blends with cotton.
(2) To produce yarns with special characteristics.

In industrial trials, an OE rotor yarn of fineness 7 tex has been successfully spun with 0.85 decitex 'trevira' at a rotor speed of 90,000 rpm with relatively few problems. The yarn tenacity attained were higher at 19 cN/tex than those, for example, of a polyester/cotton combed system ring-spun yarn. Despite the higher rotor speed of 90,000 rpm used for 100% polyester fibres, there was little fibre damage. The evenness of the yarns from the fine fibre was also found to be more uniform. However, the sharp rise in the number of neps in yarns spun from such fine fibres (0.8 decitex) indicates that there are technological constraints in processing such fine fibres.

6.3.6.1. Requirements of feed sliver to adopt in rotor spinning of microfibres. Further study reveals the fact that, in processing card slivers of microfibres on a machine set, as is customary for standard fibres, it must be taken into consideration that the slivers contained significantly more fibres. Draw-frame slivers of 1.1- and 1.6-decitex fibres possessed approximately the same sliver cohesion if they contained the same number of fibres in the cross sections (25,000). Sliver count was 4 kilotex (NM 0.25) for 1.2 decitex and 2.778 kilotex (Nm 0.36) for 1.1 decitex. In the case of draw-frame sliver of the same sliver weight (3.333 kilotex = Nm 0.30), the 30,000 fibres of 1.1 decitex produced a sliver cohesion value about 33% higher than the 20,000 fibres of 1.6 decitex. It is, therefore, primarily the number of fibres which determines the sliver cohesion. It is additionally known from the literature that with the same sliver weight, the opening role of the OE rotor spinning machine must apply about 50% greater force for 3.3 decitex fibres than for 6.7 decitex. It may be deduced from this that fine fibres undergo greater thermo mechanical stress in the opening process in the same sliver count.

6.3.6.2. Effect of opening roll. With heavy-feed slivers, if the microfibres are damaged by the relatively high friction between the fibre surface and the clothing of the opening roll, a significant part is naturally played by the nature of the opening role clothing. In preliminary trials with polyester microfibres, it was found that the rotor on some spinning heads suffered from polymer dust. The fibre taken from the spinning head that performed badly exhibited distinct damage to the fibre surface. By interchanging the spinning elements, it became clear that the opening rolls were causing the damage. Possible reasons for the differences among the different types of opening rolls are the differing wear characteristics of the surface of the roll clothing during use or difference in nickel or diamond coating.

6.3.6.3. The twist index. The slivers containing the large number of fibres in the cross section can cause problems in fibre separation. On the other hand, advantages may be expected with respect to inter-fibre friction due to the reduction of amount of twist that is required. Without using a twist-blocking device, it was possible to reduce the twist index for a yarn count of 20 tex (NM50) from 120 to 50. The rate of yarn delivery increased from around 80 to 200 m/min. A decline in tenacity started at TM 60 and yarn evenness deteriorated at TM 70. There was a very distinct sudden rise in number of thick places between TM 80 and TM 70. There is generally a beneficial effect on handle and visual appearance by reducing the difference in twist occurring with declining twist index TM. This difference between twist imparted by the machine and the twist measured in the laboratory is almost zero with ring-spun yarns, but with open-end rotor yarns, it is regarded as an index of random orientation of fibres. In product development using microfibres, a reduction of twist index should always be considered.

6.3.6.4. Rotor and delivery speeds. Ernst and Yarns [79] from the study on OE spinning of microfibres revealed that the economics of spinning process is important for fine rotor yarns. On the other hand, we require high rotor speeds on the other low-twist factors that are essential for high-delivery speeds. Extensive tests carried out using different speeds and diameters indicated the following performance limits:

(1) 100% polyester, maximum rotor speed 90,000 rpm.
(2) 100% viscose, 100% acrylic and blends, maximum rotar speed 100,000 rpm.

The experiment also revealed the influence of twist factors on yarn strength; lower twist factors demand low residual trash content in blends of microfibres with cotton. Reduction of twist factor below the optimum level increases the end breakages. Acrylic and viscose yarns were successfully spun on rotor spinning machine at rotor speeds of 120,000 rpm and more.

6.3.6.5. Polyester microfibres in blends with cotton. Apart from the basic information gained from spinning rotor yarns in 100% microfibres, a question that repeatedly arises is whether such yarns can ever attain market significance. The advantages of staple man-made fibres have always been in their facility for bending with natural fibres. This raises the question whether blends of microfibres with cotton possess advantage over fibres of standard linear density. Experiments were conducted using polyester linear densities of 0.85, 1.1 and 1.3 decitex blended with cotton in a 50:50 ratio. It was possible to spin all three linear densities at rotor speeds up to 120,000 rpm. The yarn tenacity was governed largely by single-fibre tenacity, which was highest with 1.1 decitex. Above 115,000 rpm, there was a greater tendency to neppiness when using microfibres at the same time, the number of thin places was significantly lower than with other fibres. With finer yarns, the advantage held by microfibre in respect of thin places was even more pronounced. Spinning results for blends with cotton were very strongly influenced by the grade of the natural fibre employed. By comparison, using microfibres with good cotton grade caused only an extremely small increase in the number of neps; the same applies to thick and thin places. Experiments conducted jointly with Rieter spinning systems were aimed at developing knitted and woven products incorporating 0.85-decitex microfibres. The blend ratio adopted was 50:50 and 67:33 polyester/cotton; both combed and carded cotton was used. In spinning fibre blend containing microfibre, the finest possible cotton should be used. The tenacity and IPI indices reacted less to the difference between carded and combed cotton than to the blend ratio and the yarn fineness. Using the yarns described that knitted fabrics were produced (interlock and single jersey) and woven fabrics with OE rotor yarns with warp and weft and with OE rotor yarn as weft on a filament yarn warp. The pilling behaviour was rather unsatisfactory in all the fabrics. These studies reveal that microfibres can be spun into very fine yarns on modern OE rotar spinning machines – with rotor of appropriately small dimensions – at high speeds. However, perfect condition of spinning elements, in particular the opening roll, is essential. Opening roll with a relatively heavy nickel coating is recommended.

6.3.7 Spinning microfibre yarns on the Murata jet system

Murata Machinery Limited [72] from their trials with microfibres on MJS system explored the advantages of using fine-denier fibres over the conventional ring and OE spinning systems. In recent years, fine-denier acrylic, rayon and polyester have been developed by synthetic-fibre manufacturers for use in high-quality consumer (apparel) products. However, carding is more difficult in the manufacturing process for fine-denier yarn, and drafting becomes more difficult in the ring-spinning process. Also, if the spindle speed is increased, fibres are easily damaged and, owing to this reason, the productivity is lower. As denier becomes finer, in the case of Murata jet spinning, the spinning speed increases and as more number of fibres are found in the cross section with increased contact area between them. The twist propagation of false twist generated by the spinning nozzle becomes faster. All these factors promote a faster spinning. The spinning tension in MJS system can be controlled under 30 g preventing damage to the microfibres even at higher speeds. On the other hand, when processing microfibres on ring spinning, yarn spins the traveller, and the tension increases with spindle rotation resulting in fibre damage. With OE spinning machines, the draft is performed by combing, and this combing can also damage the microfibres. For these reasons, MJS system is best suited to spin microfibres.

Yarn quality

Denier value becomes smaller, yarn strength increases, but the fibre itself becomes weaker. The yarn evenness generally improves, the number of thin places decreases, whereas neps and thick places increase. To improve the same, the static and dynamic coefficient must be reduced. Hairiness decreases as the denier becomes finer. Considering the important yarn strength, yarn evenness and yarn hairiness becomes clear that the yarn quality improves.

Fabric quality

As the denier of the fibre becomes finer, the cloth becomes softer and smoother. The cloth bulkiness in case of knitting fabrics increases as denier becomes finer, but with woven fabrics, vice versa. In general, finer denier means more trouble with pilling. However, generation of pilling varies greatly depending on the composition of the yarn. It was found that with ring-spun and MJS yarns, the finer the denier, the worse the pilling. In case of OE spun yarn, regardless of weather, the denier is fine or coarse, pilling is much worse than ring-spun or MJS yarn. For products where pilling is an important factor, OE spun-yarn cannot be used.

6.3.8 Air-jet spinning of microfibres

In an experimental work on air-jet spinning (a type of OE spinning), as reported by Rajamanickam et al. [80], microdenier polyester/cotton blended yarns were spun to study the interaction of first and second nozzle pressures on the properties of spun yarn. The results show that the first and second nozzle pressures interact with each other to determine yarn strength. This is because both the number of wrapper fibres and the length of wrappings formed at a particular first nozzle pressure depend on the level of second nozzle pressure and vice versa. There is an optimum number of wrapper fibres and wrapping lengths that yields the maximum yarn strength, and this optimum level can be obtained at several different nozzle pressure combinations because the first and second nozzles interact with each other. However, it would be advantageous to use the lowest of these nozzle pressure combinations because of significant savings in energy costs.

Rajamanickam et al. [80], investigated the effect of yarn delivery speed, first nozzle pressure and blend ratio on the hairiness profile microdenier polyester/cotton blended air-jet spun yarns. The results showed that yarn hairiness increased with increased yarn delivery speed and first nozzle pressure, but decreased with increasing amounts of polyester in the yarn. Also, the three factors (yarn delivery speed, first nozzle pressure and blend ratio) interact with one another in determining yarn hairiness.

6.3.9 Compact spinning of microfibres

A new impulse in the field of ring spinning technology is offered by compact-condensed spinning. The spinning triangle that occurs while the yarn is formed is the cause of many fibres leaving the drafted roving or being partly spun into the yarn with one end only. This causes greater waste of fibres, low exploitation of fibre tenacity in yarn, poorer appearance and greater hairiness in spun yarn. The newest research in the field of ring spinning has shown that modification of three-cylinder drafting equipment with two aprons in a region after front-drafting rollers enables ring spinning to proceed with a minimised spinning triangle or even without it all. This is called as compact or condensed spinning [82]. In the context of the cooperative relationship between Rieter machine works limited, Switzerland and Reliance Industries limited, India, the processing properties of a Reliance micro polyester quality has been examined in a joint research project with Reutlingan technical college. Schenek and Schwippl [88] report from the trials on compact spinning of micro polyester fibres in Rieter COM4 machines that COM4 yarn displays better yarn uniformity indicating the necessity of fibre compacting that has a positive influence on fibre orientation. This is shown in Figure 50. Fibre Compacting, thus, has a positive influence on fibre orientation. The number of fibres in cross section or the twist factor should, therefore, be reduced to meet high-quality standards.

Regarding imperfections, compacting reduced thin and thick places in the yarn by up to 20%. No differences were apparent due to compacting with regard to neps. The COM4 spinning system with good drafting behaviour and fibre integration in the spinning integration offers interesting potential for the end product. Despite the large number of fibres in the fibre, cross section's increase in tenacity of 1 cN/tex compared with conventional ring spun yarn was achieved as a result of better fibre integration in the COM4 spinning system. COM4 yarns display no reduction in mean tenacity as a result of reducing the twist factor

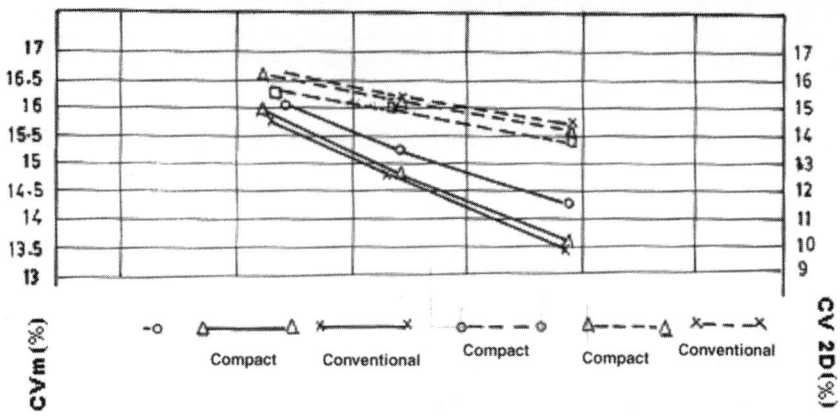

Figure 50. Yarn irregularity of micro polyester compact yarn.

of $\alpha m = 108$ or $\alpha m = 97$, respectively. This would, therefore, be an option for increasing output without the loss of quality.

Role of microfibres in compacting enhances evenness because of more number of fibres in cross section, thereby reducing thick and thin places. Also, the same yarn tenacity, as obtained with ring spinning, can be achieved in microfibres with less twist multipliers as we have a large surface area of fibres for better cohesion, and the higher number of fibres in cross section helps in reduction of TM. The low hairiness of COM4 micro-polyester yarns is clearly noticed compared with conventional ring-spun yarn measured according to uster H and Zweigle 1 + 2 mm. In these Zweigle S3 values, a small number of hairs per metre of yarn are seen. In subsequent winding process, lower hairiness has a generally positive effect on nep counts due to less fibre sloughing. The COM4 yarns also possess a better abrasion resistance.

Micro lyocell is notably successful in compacting up to a yarn count finer than 30 tex. Hairiness was reduced substantially in comparison with conventional ring spinning technology with a very fine yarn of approximately 10 tex. At the same time, tenacity increased by up to 3.5 cN/tex as reported from Rieter LINK48 [71].

6.4 Typical processing line for a synthetic fibre

A typical processing machinery is discussed, taking the example of acrylic. Reliance Industries, India, has introduced Recrylic™ Microdenier acrylic fibre (0.9 D). Garments, especially sportswear, made from Recrylic™ micro have been found to be soft, comfortable and with good drape. The fibre, as claimed by the company, has to be spun carefully for achieving good results.

Pre-opening

Pre-opening is not required for 100% microdenier acrylic staple fibre (ASF), but it is better to have the bale wires opened at least a day before the processing to relax the fibre at a temperature of 32°C and relative humidity (RH) 65%.

Over spray

It should be slightly lower than the quantity used for 1.2 D fibre (as the card-loading/roller-lapping tendency will be slightly higher). If tinting colour is used, it has to be of lighter concentration.

Blow room (Temperature 32°C/RH 65%)

Optimal beating point: 1
Sequence

(1) One bale opener
(2) BR inclined lattice
(3) Kirshner beater (<400 rpm)
(4) (ERM, RK, KB/AFC, STEP)

Lap weight/metre

About 20–30 g lower than 1.2 D (applicable for chute feed also). Beaters such as porcupine (RN), monocylinder, saw-toothed beaters can be bypassed. Disc beater is more suitable as it prevents the rupture of the fibres. Depending upon the initial cohesion level of the fibre, blender opening, evener roll-lattice setting and fine-beater speed can be optimised.

A lap weight of 350–380 g/m would be ideal.

Carding (Temperature 32°C/RH 58%)

If production balancing permits, hanks can be made finer by 5%, and the delivery speed can be reduced by another 5%. The recommended high production card delivery speed is 80–100 m/min. For less neps and yarn fault, the card wire condition should be very good.

Draw frame

Top roller loading should be uniform.

Simplex (Temperature 32°C/RH 55%)

Twist multiplier can be reduced by 3%–6% (as compared to regular-fibre TM). Optimum simplex TM can be maintained at 0.68–0.72. The condition of false twisters should be good.

Ring frame (Temperature 28°C–32°C/RH 47%–51%)

Though there is a reduction in speed at carding, productivity can be compensated by reducing TM at ring frame.

In the ring frame, TM can be reduced to 2.65–2.75 for recrylic micro as against 3.0 TM used for normal acrylic (for 1/20 Ne, hosiery application). Low TM will also add extra softness to the fabric.

Tips for enhanced performance

(1) Top roller condition should be smooth and free from knife/hook damages.
(2) Top roller cleaning should be done once a week.

(3) Traveller-EM1-UDR/M1 UDR type may be used. Shorter traveller-changing schedule may be followed.
(4) Pneumatic suction should be checked.

Post-spinning (Temperature 32°C/RH 65%)

Splicer strength should be ascertained by fine-tuning UNIBOX splicers/Auto coner splicers (MESDAN, Italy). Splice strength should be around 85% of the normal yarn strength.

Applications

The applications of different types of recrylic yarns and the popular counts are given in Table 11.

Chenille yarn

Recrylic microdenier acrylic fibre creates superior chenille yarn, which can be used both in dress material and home furnishings. Dry-spun acrylic microdenier chenille yarn can enhance the performance of textile fabrics used in upholstery like sofa and automobile seats because of its unusual softness and exceptional colour fastness. Chenille yarn having anti-microbial properties can increase the value proposition for this application.

6.5 Migration behaviour of microfibre in short staple spun yarns

The structural study of spun yarns includes the characterisation of fibre migration, which has a decisive influence on the mechanical and physical properties of the yarns. Fibre migration, apart from being influenced by constituent fibre properties, is also influenced by the spinning systems adopted with specific fibre-accumulation mechanism and the process conditions in yarn formation. Among the modern techniques, OE rotor and friction-spinning systems have drawn much attention. When migration is considered, ring-spun yarn exhibits higher migration, followed by rotor-spun yarn and friction-spun yarn with the least [84]. Migration and packing density are two independent factors that contribute to yarn strength. Higher value of these factors corresponds with higher yarn braking tenacity realisation.

The recent development of compact spinning technology for staple yarns, in which the size of spinning triangle has significantly gone down, has reduced the migratory characteristics of fibres to a certain extent in constituent yarns due to relative changes in the tension gradient of fibre strand at the point of yarn formation. In those yarns, the strength gain is obtained mainly through the favourable change in the packing density of yarn. In general, it could be concluded that the yarn structure depends on the one side on spinning technologies and, on the other side, on processing conditions, and this structural difference in staple yarns leads to different yarn properties. Thus, it is very important to understand yarn structure and its effects on physical properties of yarn, because each kind of yarn manufactured by

Table 11.   Usage of microfibre yarn for various applications.

Yarn for various applications	Popular counts
Hosiery and chenille	1/20,1/30,1/24,1/40 Ne
Weaving	1/20,1/40,2/40,2/60 Ne

specific spinning system method exhibits unique properties. The relative fibre movement at the point of yarn formation and the resultant position of fibres in the yarn structure are described as fibre migration. The migration behaviour of a fibre is affected significantly by the inherent properties of constituent fibre like fibre length, fibre fineness, crimp and cross-sectional shape and the inherent characteristics of adopted processing systems. Of the widely used systems such as ring, rotor and friction, ring-spun yarn exhibits the highest fibre migration followed by rotor- and friction-spun yarns on the basis of spinning tension and its variation.

Fibre migration of regular ring-spun and OE microdenier modal fibre yarns has been studied by Ramakrishnan et al. [85] using the tracer fibre technique for the different migration parameters. The yarn counts used were 20 Ne (tex). Identical twist levels were used for ring and rotor yarns. In addition, a higher twist level was used for OE yarn, as it is required for deriving equivalent strength of ring yarns. The study revealed a reduction in mean fibre position to the tune of 22.6% and 37.2% and reduction in RMS deviation to the tune of 5% and 6%, respectively for OE-LT & OE-HT yarn compared with ring yarn, which is statistically insignificant. There is an increase in migration intensity to the tune of 30.3% and 58.2%, respectively for OE-LT & OE-HT compared with the ring yarn, and so the equivalent migration intensity and migration factor. This may be due to the variation in stress, strain values of fine-denier fibre, and also the flexural rigidity is less for the fine-denier fibres compared with the coarser fibres. From the present study and other previous works, it could be inferred that mean fibre position is mostly affected by the spinning system used; RMS deviation is affected mostly by the process variables in the spinning system used and migration intensity is affected by the fineness of the fibre being used apart form twist intensity.

6.6 Warping

In case of warping microfibre yarns compared with standard higher denier yarns, there are two major factors that change.

The first is the smaller amount of force that is required to stretch and/or break a microfibre filament. For example, a 2-denier filament in a fully oriented yarn can take forces of 10 to 12 g before breaking. Whereas a 1-denier filament in a partially oriented yarn going into a draw warping microfibre process can take only 1 g of force or less before stretching beyond recovery and subsequently breaking during the preceding draw zone.

The second fundamental change is the higher surface to volume ratio of filaments and this means that more finish is required to get some degree of protection and lubrication. Most microfibre yarns are draw warped, instead of warped. However, low-denier yarns with 1 decitex per filament are often used as standard warping, and the process is similar to that required for microfibre yarns. For these operations, the creel structure is the same as the packages are large in size (450 mm diameter), which requires creels that are much larger, or either longer or higher. The best solution is often a two-storey creel the effect of which is to increase the yarn tension due to contact with guides, air drags and yarn break angles. The yarn is led into the most critical element in warping, that is, the tension device. The basic principle of tension device is that it never takes away tension but only adds tension. Also, it is important for a tension device to mechanically respond and compensate for package tension changes during high-speed warping. Therefore, what is recommended for microfibre yarns is that it should be simple and provide absolute minimum of tension. To accomplish tension profile at a low level with microfibre yarn, it is also necessary to pay attention to all surfaces that come in contact with the yarns. For instance, the eyelets used in the tension device, eye boards and eye bars in the creel must be made of a low-friction material with a matt surface.

Another important element in the warping of microfibre yarn is the reeds. The reed blades must be absolutely free of any snags or burrs. In general, standard straight reeds used in most warping operations will not work with microfibre yarns, and they call for specialised manufacture [86].

6.7 Sizing

Voswinckel [87] reports from the study on sizing of microfilament yarns that sizing of filament yarns usually have three times as many single filaments with double surface and three times as many contact points than standard yarns. As an effect, specific conditions have to be met when using microfibres during warping, warp sizing and assembling to avoid faults and to work efficiently. The thread surface of microfibre yarn is tending to 15% larger, and the total surface of the individual capillaries is approximately twice as large with the interspaced volume still being the same. The fine single fibres are more delicate if exposed to mechanical influences like frictions. They show a clear propensity of getting electrostatically charged. Owing to their small mass, they tend to be heated faster and thread absorbs more sizing agents with the thread surface being larger. The tensile strength of threads is slightly higher.

Practical applications show that due to large net surface of the microfibre, smoothed yarn absorbs 10% more sizing agents and textured yarns 15% more sizing agents provided sizing as the same recipe. Sizing could be reduced by cutting down the concentration. This would reduce the viscosity and improve distribution of the sizing agent without lowering adhesion of the thread since the capillary spaces are considerably smaller, which improves the adhesion. The squeezing pressure must be adapted to the speed so as to achieve regular sizing. The microfibres carry approximately 10% more Aviv age in the sizing bath. For this reason, apart from skimming off the Aviv age that deposits in the pretruf, it is recommended to provide for an oil separator either as centrifugal or as micro-filter destined to continuously separate the spinning oil in bypass circulation. Dry out of the sizing agent is the most critical stage after the application of sizing agent. As a result of the larger surface and the small mass of the individual filament, the microfibre can be heated faster as dry out also occurs faster. For microfibres, the drying temperature was cut down so that the drying capacity was reduced accordingly. During sizing process, the overall stretch of microfibre yarns must be maintained during all operation stages. In principle, sizing of microfibre yarns does not require new processing or machine techniques but needs a critical optimisation at every stage.

6.8 Weaving

Nau [88] reveals from the study on requirements of weaving microfibre yarns that the new generation of fibres naturally has its influence on the development of weaving machine, including its accessories that come into direct contact with yarns made with its fibre. The harness with the droppers of the warp stop motion; the reed and the healds come into intimate contact with the microfibre yarns. The surfaces of these items, therefore, need to have low roughness. For weaving microfibre weft, reeds have been developed that are fitted with specially shaped reed wires. As fabrics woven from microfibre yarns are often densely constructed, beat up must also be relatively severe. If the reed wires have sharp edges, they can easily cut through the fine individual filaments and, thus, damage the yarn.

For this application, reed wires are therefore cut from flat steel, and their edges are then rounded. Reed wires with a slight crown on the narrow edges have proved very useful. Chrome steel and Nickel chromium steel are used, and the reed must also be corrosion resistant depending upon the yarn being woven and the method of weft insertion. Healds makers are, likewise, concerned about microfibre yarns and have been providing surface treatments or plasma coatings to cater to the requirements of weaving microfibre yarns. Weaving machines need no modification for weaving microfibre yarns, and such yarns may be woven on machines designed for silk and silk-like fabrics and those for weaving fine filament yarns. In weaving microfibre and micro-filament yarns, the temple cylinders must be compatible with the needs of the fine fabrics. Temple makers also recommend types with plastic rings, with fine pins, single rings or twin rings running separately on the same ellipse or possibly with auxiliary cylinders.

6.9 Knitting

The advantages of microfibres have been well exploited through the knitting process. The lower denier fineness of microfibres proves to be physiologically advantageous especially in wear situation where heavy and copious perspiration exists. In these circumstances, the microfibre textiles show better moisture transport and ensure better moisture control than the other construction parameters of comparative textile from conventional fineness above 1 denier in the microclimate near the skin. The reason for the better physiological isolation of the microfibre textiles in these wear situations can be attributed to the higher absorption potential of the fibre surface enhanced by the fibre fineness as well as a better capillary effect during the transport of liquid perspiration. Knitted fabrics have the advantage of flexibility. When a sport or activity requires a wide range of body motion, highly elastic knits offer a number of practical benefits. Knitted fabric possesses stretch providing full freedom of movement and, in particular, has two important functions to perform, namely to provide free movement and transmission of body heat to the next textile layer in the clothing system.

In an investigation on properties of knitted fabrics made from viscose microdenier fibres reported by Ramakrishnan et al. [89], the properties of the knitted fabrics made out of microdenier (<1 denier) and normal denier (>1 denier) viscose yarns were compared. Results highlighted that microfibre fabrics show excellent drapeability. The moisture transmission properties like wicking and water absorbency show better results than viscose normal denier fabrics and other synthetic microfibres. The microfibre-knitted fabric was dimensionally more stable when compared with the normal-denier knitted fabric because of less loop-shape deformation and characterised by better stitch density and tightness factor.

Karolia and Paradkar [90] observed knitted double-jersey fabrics of polyester microfibre and non-microfibre fabrics for growth and elastic recovery by load application, tensile strength and percent extension, flexing behaviour and abrasion resistance of fabrics. The study revealed that microfibre fabrics were superior in physical comfort than non-microfibre fabrics. Microdenier fibres have excellent flexibility and yarns with better regularity and elongation that contribute for perfect knittability ensuring knitted fabrics with better softness, drape, dimensional stability and wicking, thus ensuring better mechanical and comfort properties. The hairiness of the microfibre yarns is very low; this, in turn, creates a low lint-shedding propensity; and it will generate lesser fly during knitting. These two aspects of lint shedding and fibre fly are crucial for the improved efficiency of the knitting machine. Matic-Leigh [91] observed from their study on bilayered knitted fabrics with polypropylene inner and cotton outer were seen to provide better comfort for an ideal sportswear.

Figure 51. Loop formation for normal fibres on knitted surface.

Microfibre polyester knitted fabrics have demonstrated superior growth and elastic recovery properties besides having a good tensile strength, extension, flexing behaviour and abrasion resistance. Microfibre blends with natural fibres were observed to enhance production speed in knitting besides improving softness drape and better dimensional stability of knitted fabrics.

In a study on properties of knitted fabrics made from microdenier polyester fibres, Ramakrishnan and Mukhopadhyay [92] reported that the microdenier fabrics show superior properties when compared with normal-denier fabrics in various aspects of physical and dimensional behaviour. The microfibre fabrics were characterised by high drapeability, acceptable spirality and excellent moisture-transmission properties such as drying rate, total absorbency, wicking rate, drop absorbency and water absorbency. The microfibre knitted fabric was dimensionally more stable when compared with that of normal-denier knitted fabric because of less loop-shape deformation and characterised by lesser lint-shedding propensity. The superior properties of microfibre fabric can be conveniently utilised to explore and optimise new products for apparel and sports wear. Scanning electron micrographs (Figures 51 and 52) clearly indicated finer and lesser number of loops for the microfibre-knitted fabric. The finer denier microfibre being more pliable in nature gets better incorporated in the yarn resulting in neat surface appearance. The surface thus develops a smoother look though the same moisture absorption capacity was retained. The loop for the cotton knit was characterised by the ribbon-like convolutions on the surface, which was the characteristic of cotton fibre. The overall surface as characterised by Figure 53, and Figure 54 showed the uniform surface of the microfibre-knitted fabric. The surface was viewed from a different angle that gave a glimpse of the overall surface appearance of the three knitted structures under consideration. The lesser number of loops resulted in a more coherent structure of the microfibre-knitted structure. The thicker protrusion of the normal and cotton-knitted structure imparts an overall rougher appearance.

In textile structures, the spaces between fibres effectively form capillaries. The closer the fibres in yarns the smaller the apparent diameter and the more readily wicking can occur. Fibre properties such as diameter, cross section, crimp and stiffness, all play a role in capillary formation. For example, microfibres packed together very tightly form narrow

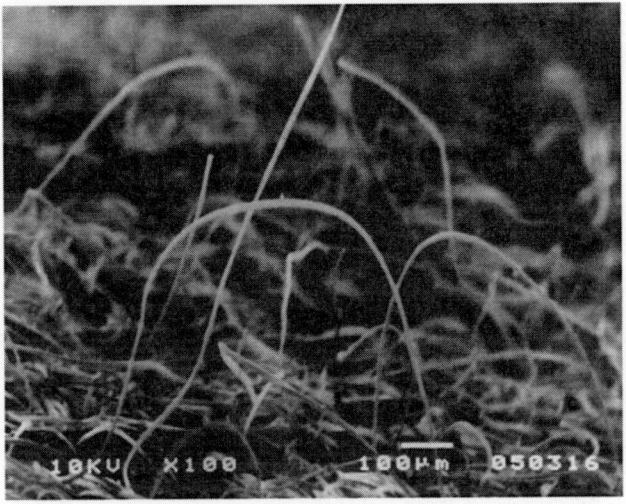

Figure 52. Loop formation for microfibres on knitted surface.

capillaries, enabling fabrics made from microfibres to pick up sweat more easily than those from conventional fibres.

The sportswear developed by Sportswool Pro[TM] [93] is a superfine merino wool sportswear on the inner face, and the mean diameter is 18.3–18.5 μm, ensuring skin comfort in active conditions where the skin is moist and highly sensitive. The outer face can be one of the several different polyesters, depending on the aesthetic requirement of the end product. The surface energy of machine was treated wool is lower than that of most polyester. The specific fabric-finishing processes used with Sportswool Pro[TM] increase the surface energy of the wool slightly, but greatly increase that of polyester, enhancing the difference in surface energy between the two faces. The combination of both diameter and surface energy differences results in a strongly positive wicking gradient between the inside and outside of the fabric. Sweat picked up from the skin by Sportswool Pro[TM] quickly

Figure 53. Surface of normal-fibre knitted material.

Figure 54. Surface of microfibre knitted material.

pulled through to the outside face by the wicking gradient, where it then increases considerably in area. By increasing the area of sweat exposed to the air moving past the wearer, the sportswear maximises the opportunity for evaporation and efficient body cooling as reported from Sportswool Pro^{TM}.

Microfibre bilayered were developed by Ramakrishnan and Nandhini [94] for enhancing the wearer comfort. Three different types of bilayer socks were produced using micro-polyester, (both filament and staple), micro-polypropylene in the inner layer and micro-modal in the outer layer, as shown in Table 12.

These were compared with the regular socks product for transmission properties such as air and moisture. The results indicate a substantial improvement in comfort levels by the use of high-performance microfibres as indicated by the moisture vapour transmission rate in Table 13. From this table, it is understood that micro-polyester filament yarn fabrics have the highest moisture transmission rate followed by micro-polyester staple yarn fabrics. The results show that micro-polyester transfers the sweat quickly than nylon and polypropylene fabrics.

Sule et al. [95] from an investigation describe the main prerequisites for sportswear for hot and humid climate conditions in India as (a) protection from heat and (b) rapid dissipation of sweat. No sportswear made from any single fibre or blends of different fibres can make an ideal sportswear. It was recently recognised that it is necessary to have a multilayer sportswear. Such a sportswear should have an inner layer of a hydrophobic fabric that acts as a wicking layer transporting sweat to a hydrophilic outer layer of the sportswear that absorbs the sweat and spreads it rapidly to evaporate it producing a cooling effect; in

Table 12.   Details of Socks using microfibres.

Type	Inner layer	Outer layer
Socks A	Micro polypropylene (filament)	Micromodal
Socks B	Micro polyester (filament)	Micromodal
Socks C	Classic nylon	Micromodal
Socks D	Micro polyester (staple)	Micromodal

Table 13.   Moisture Vapour transmission rate of socks.

Drying threshold (%)	1.05	1.07
Speed of drying (1/t)	5.3	3.9
Speed of capillary rise (cm^2/s)	0.5	0.41

addition, a sportswear should also be soft to touch. Very often the wicking action is faster than evaporation in which case the perspiration is transferred back accumulating around the skin making active sportsman very uncomfortable. Hence a proper balance in function of the inner and outer layers is required in designing sportswear. The psychological aspect can also be taken into account by colouring the upper layer blue if protection is required from heat. The three main prerequisites of a good sportswear or leisurewear are form fitting and shape retention, protection, comfort, etc. It is not possible to achieve the desirable attributes of a functional sportswear such as optimum and moisture regulation, good air and water vapour permeability, dimensionally stable, etc., using a simple structure of any single fibre or blend. The ideal sportswear should be a multilayered structure. The use of superfine or microfibre yarn enables the production of dense fabrics leading to capillary action that gives the best wicking property. The twin layer structure in sportswear is a concept totally different from blending of fibres. The object of the layer next to the skin is to wick away the perspiration rapidly to the outer layer, which absorbs and dissipates it rapidly to the atmosphere by evaporation.

6.9.1 Properties of knitted fabrics made from microdenier fibres

Table 14 illustrates the comparative properties of fabrics made of different kinds of acrylic fibres. The fabrics made out of microdenier fibres are superior in many cases as compared to that of fabrics produced by conventional fibres.

As can be seen from the results, the microdenier improves draping properties. The soft towel of microfibre is represented by the high value of the compression recovery test. This is confirmed by good air permeability measures. In this particular case, it was observed that in spite of thin filaments, the pill formation resistance was substantially different from the standard acrylic. The original end use sector of microfibre was active sportswear with a superfine microfibre yarn, where it was possible to construct fabrics that are so fine that they were water and wind proof. For the knitting industry, the main end use sector should be warp-knitted and surf-knitted fabrics that embrace both traditional functional active sportswear and highly fashionable leisurewear as observed by Hauer [96].

7. Techniques to improve wetting of microfibres: hydrolysis

Microfibre fabrics are generally lightweight, resilient or resist wrinkling, have a luxurious drape and body, retain shape and resist pilling. Also, they are relatively strong and durable in relation to other fabrics of similar weight. The fineness of the microfibres allows many fibres to pack together very tightly. With many more fine fibres required to form a yarn, greater fibre surface area results making deeper, richer and brighter colours possible. The details of the dyeing-process developments and care for the microfibres have been discussed in the dyeing section.

The higher packing density of microfibres enables the formation of garments requiring wind resistance and water repellency. Yet, the spaces between the yarns allow the fabric to breathe and wick body moisture away from the body. When comparing two similar fabrics,

Table 14.   Properties of microdenier knitted fabrics.

Properties	Unit	0.8 decitex	1.3 decitex
Compactness	Tex/cm	14	13
Weight	g/m^2	213	208
Thickness	mm	1.08	1.10
Voluminosity	cm/g	6.1	5.3
Flexural rigidity	mg/cm	26	2.6
Flexural length	cm	0.95	1.11
Compression recovery	%	49	45
Wear resistance	Cycles	6400	5000
Air permeability	L/cm/min	6.3	6.5
Pilling	0–9	5	6
Insulating capacity	min	26.9	28.1
Water absorption	%	42	32

one made from a conventional fibre and one from a microfibre, generally the microfibre fabric will be more breathable and more comfortable to wear. Microfibres seem to be less 'clammy' in warm weather than conventional synthetics.

The subsequent paragraphs discuss the reaction of microfibres to alkaline and enzymatic hydrolysis and the effect of chemical splitting of microfibres.

7.1 Alkaline hydrolysis

The effect of hydrolysis by aqueous sodium hydroxide on regular and microdenier polyester was investigated by Hsieh et al. [97]. They observed significant changes in surface-wetting properties and pore structure of regular and microdenier PET fabrics. Two-hour hydrolysis at a 3N NaOH concentration reduces fabric weight and thickness and increased the overall porosity. The extent of these changes increased with increasing hydrolysis temperatures. The effects of alkaline hydrolysis on the physical properties of these two fabrics were, however, found to be clearly different. The reduction in fabric thickness is more significant for the regular polyester fabrics, whereas pore enlargement predominates in the microdenier polyester fabrics. Hydrolysis reduces the thickness of the regular PET fabric significantly, while enlarging pore sizes evenly. For microdenier PET, fabric thickness is reduced to a lesser extent, but the enlargement of the bigger pores is greater than that for the regular PET fabrics.

From the hydrolysis of the regular and microdenier PET, the results of this study show that varying aqueous alkaline hydrolysis conditions causes different changes in regular and microdenier fabrics in terms of their wetting properties and pore structures. Increasing hydrolysis temperature (3N NaOH) is more effective than hydrolysis time in producing effective pore structure for water transport and retention. Hydrolysis (3N NaOH at 55°C) as short as 10 min can significantly improve surface water wettability without changing surface topography. Improved water retention in regular and microdenier PET fabrics is primarily associated with improved wettability of the fabrics. Water retention values of hydrolysed regular and microdenier PET fabrics are positively and linearly related to their water wettability, as indicated by their water-contact angles.

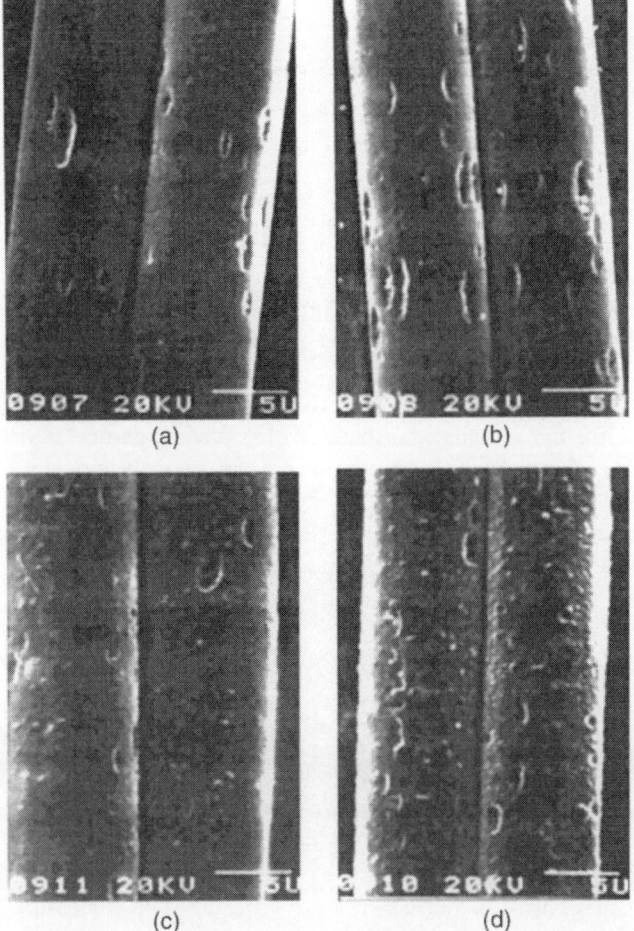

Figure 55. SEM photos of NaOH- and NaGR-treated PET microfibre fabric surfaces at similar levels of weight loss ratio: (a) 29.2% by NaOH, (b) 47.6% by NaOH, (c) 34.3% by NaGR and (d) 47/7% by NaGR (after Huh and Yoon [100]).

7.2 Enzymatic hydrolysis

Both alkaline and enzymatic hydrolysis cause more significant improvement in water wetting of the microdenier PET fabric than in its PET counterpart. Due to the large sizes of enzyme molecules [98], PET hydrolysis by lipases is considered a surface phenomenon. How hydrolysis reactions differ on the two kinds of PET fibres and the exact causes for the lower contact angles of the hydrolysed microdenier PET are not completely clear. The total fibre surface area of the microdenier fabric is higher than that of the PET by 17 and 112% at same fabric size and volume basis, respectively. Assuming similar hydrolysis, the larger number of polar groups due to the higher surface area on the hydrolysed microdenier polyester fabrics could explain their lower water wetting contact angles. It is also possible that the morphology in these fibres may be different due to the different spinning processes. These different surface morphologies may lead to different surface hydrolysis.

7.3 The effect of chemical splitting

The effect of chemical splitting on the water absorption and mechanical properties of a split-type N/P microfibre pile knit was investigated by Kim et al. [99] under various alkaline hydrolysis treatment conditions. Weight loss of the hydrolysed pile knit increased as with an increase in hydrolysis time, temperature and concentration of sodium hydroxide. Areal density decreased with increased weight loss.

Huh and Yoon [100] studied the effect of alkaline conditions on polyester microfibre fabrics. The fabrics were alcoholysed at 120°C, 140°C and 160°C in 0.5, 1.0, and 1.5% (w/w) sodium glycerolate/glycerol solutions (NaGR) and the decomposition kinetics was studied in comparison to the hydrolysis done by a 5% aqueous sodium hydroxide solution (NaOH).

The uniformly developed microcrater feature (Figure 55) of the alcoholysis may have some advantage for the softening finishing of PET fabrics, particularly PET microfibre fabrics. As opined by the authors, this feature of the alcoholysis might prevent sudden physical deterioration of PET fabrics and may improve the tactility of PET fabrics. The authors also commented that these features would increase the dispersed light reflections, thereby giving a silk-like effect to PET fabrics.

8. Microfibre dyeing

8.1 General dyeing of microfibres

Dyeing of microfibres is different from normal-denier fibres due to the increased surface area of such fibres. The subsequent sections discuss the developments in research for microfibres and also some of the special considerations for microfibre dyeing.

Nakamura et al. [101] studied the sorption and diffusion behaviour of disperse dyes on microfibres. Sorption isotherms and dyeing rates of purified disperse dyes on polyester microfibres (fineness of 0.25–1.0 denier) from water have been measured at 95°C. The isotherms were curved and followed the dual-mode sorption model: Nemst-type partitioning and Langmuir sorption, which were found to be concurrently operative. The effect of a diffusional boundary layer on the dyeing rate was found to be small under the conditions that the microfibres were dyed in the form of a bulky two-ply yarn in a well-stirred bath. Dyeing rates of a commercial dye were also measured at 110°C and 130°C. For dyeing of a 0.32-d fibre at 130°C, the amount of dye sorbed by the fibres attains a maximum value at an early stage and then decreased gradually. This phenomenon is explained in terms of the aqueous solubility of very fine dye particles.

Burkinshaw et al. [102] reported the use of 13 acid dyes on conventional and microfibre-knitted nylon 6,6 fabrics with four different dyeing methods. The dyes exhibited a faster rate of uptake, higher extent and rate of dye desorption, lower wash fastness and lower colour strength on microfabric than on a conventional one. These findings were attributed to the greater surface area of the microfibres.

It is not easy to predict the termination point of dissolution process from the weight reduction curves of the fabrics containing sea–island type microfibres. SEM serves as a useful tool for the verification of dissolution, but the method is costly and time consuming. In a study by Koh et al. [103], the dissolution of radial-type polyester microfibres was monitored by a cationic dye-staining method. The weight reduction behaviour of alkali-treated microfibre fabrics and of treated fabrics stained with cationic dye was compared. Termination of monitoring of dissolution by both methods was also confirmed by scanning electron microscopy. The researchers found the cationic dye-staining method to be a much simpler and effective method in predicting termination of dissolution process.

The effect of supercritical dyeing conditions on the morphology of polyester microfibres was studied by Drews and Jordan [104]. They have showed that supercritical dyeing has no adverse effect on the fibre structure.

Dyeing properties of a polyester taffeta made from ultrafine fibres (0.07 denier), made using sea-island-type conjugate spinning techniques, with disperse dyes were studied by Nakamura et al. [105] through an analysis of sorption isotherms and rate of dye-sorption data. This was compared with data for microfibres (0.25, 0.32 and 0.44 denier) made by the conventional melt spinning method. Physical properties of the ultrafine fibres relating to the dyeing properties are also measured. They found that sorption and diffusion behaviour of disperse dyes in polyester ultrafine fibres made by sea-island-type spinning techniques is almost the same as that of polyester microfibres made by conventional melt spinning if the fibres contain no additives.

Dieval et al. [106] studied the polyester microfibre and fibre structure by critical dissolution time. The results showed that microfibre has a structure allowing good diffusion of the dye and showed that independent of the fineness; structural differences do exist between classic polyester and microfibres.

Chen et al. [107] studied the dyeing properties of polyester microfibres and regular polyester filaments. The dyeing rates were measured for four disperse dyes, the K/S values and the colour properties of cross-sectional views of the dyed fibres. The research reveals that at 70°C, dye exhaustion of microfibres was very small, meaning that dyeing the fibres at a low initial dyeing temperature was less effective in achieving good colour effects, particularly with high-energy disperse dyes. Also, when dyeing microfibres, small disperse dyes were found to require a low dyeing temperature, and bulky dyes require a higher dyeing temperature. Microfibres had a lower initial dye exhaustion temperature than regular fibres. The effective end dyeing temperature for polyester microfibres was found to be around 125°C for small molecular structure disperse dyes, but the effective dyeing temperature was around 130°C for large-molecular-structure disperse dyes. The heat-setting temperature for the lowest dyeability of polyester microfibres was usually 15°C lower than that of the regular fibres in this experiment. The surface reflectivity of the microfibres was usually higher than the regular fibres, and this led to lower K/S values than those of the regular fibres.

8.2 Special considerations for microfibre dyeing

Dyeing of microfibres is not very different from normal fibres except the fact that microfibres require more dyestuff to reach the same shade. Getting dark shades is also very difficult.

8.2.1 Reduced depth of shade

Kobsa et al. [108] of Du Pont have used an optical ray trace model to do multifilament scattering in a yarn bundle. The above researchers at Du Pont applied the method to a series of schematic yarn bundles of the same total denier and dye concentration, but with different numbers of filaments in each bundle (Figure 57).

The results provide insight into the fundamental interaction of light with a yarn bundle. As commented by the authors, in addition to reducing dye yield, reducing denier per filament increased back scattering of light from the fabric. Consequently, the fabric appears brighter, less dark. This effect is generally confused with the real effect of denier per filament on

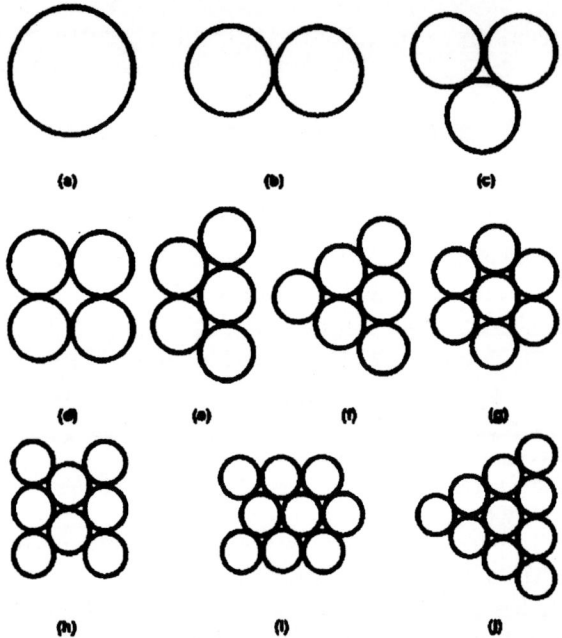

Figure 56. Schematic yarn bundles with the same assumed total denier (10.0) but different numbers of filaments. The deniers per filament are (a) 10.0, (b) 5.0, (c) 3.33, (d) 2.50, (e) 2.0, (f) 1.67, (g) 1.43, (h) 1.25, (i) 1.11 and (j) 1.00 (after Kobsa et al. [108]).

dye yield, even though it is caused by a mechanism that does not involve the dye at all. The researchers have modelled the phenomenon that predicts the expected reduced dye yield with decreasing denier per filament. The data agree well with the empirical relationship that the amount of dye required to achieve a given depth of shade varies inversely as the square root of the denier per filament. The reduced dye yield of the smaller filament bundles is a result of the reduced path length of light in these filaments. Reducing denier per filament increased back-scattering from the fabric, in addition to reducing dye yield.

The researchers plotted absorption coefficient, required to achieve the same absorbance as the single filament using $\alpha = 0.01$ μm, against the inverse of the square root of the denier per filament for each of the yarn bundles in Figure 57. The linear relationship suggests that the amount of dye required to achieve a given depth of shade varies inversely as the square root of the denier per filament.

To better understand the fundamental relationship between appearance and fibre cross-sectional geometry, Kobsa et al. [109] developed a ray trace lustre prediction computer program [110] to simulate the light scattering from a fibre. The researchers explained the reduced dye yield of microdenier yarns with the lustre prediction computer model. Results from the model demonstrated that the reduced dye yield is a necessary consequence of the reduced path length of light inside the filaments.

8.2.2 Stagnant solution layer

The effect of a stagnant solution layer surrounding the fibres on the dyeing rate is greater when dyeing microfibres [111, 112]. Therefore, an uneven flow rate of dye liquor through the fibre assemblies is apt to cause uneven dyeing. The dye absorbed in the fibres will

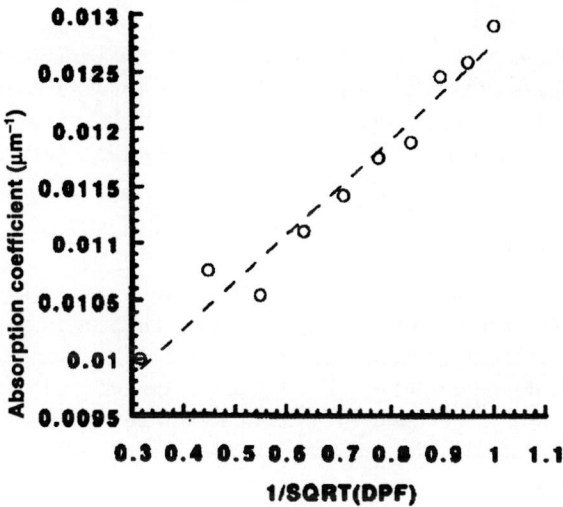

Figure 57. The absorption coefficient (μm-1), proportional to dye concentration, required for each yarn bundle in Figure 56b to j to achieve the same absorbance as the single filament with $\alpha = 0.01$ μm^{-1} (after Kobsa et al. [108]).

diffuse to their surfaces during a heat-setting process, which results in poor wet fastness of dyed fabrics since microfibres normally contain greater amounts of dye than regular ones. The visual shade depth of textiles decreased with decreasing fibre radius [113]. Studies on the relation between the amount of dye on the fibre and the visual colour depth, and the end-use fastness properties of the microfibres have been studied by Leadbetter and Dervan [114].

8.2.3 Thermomigration

Even though loosely bound dyes on the fibre surface are completely removed through reduction clearing after dyeing, subsequent heatsetting promotes dye migration from the fibre interior to the fibre surface, leading to lower washfastness. This phenomenon is known as thermomigration, which is the diffusion of dye as a result of the breakage of the interaction between dye and fibre by the increased thermal motion of both the dye molecule and the polymer chain at elevated temperatures. This thermomigration becomes more severe as the denier of fibre is smaller, and it is regarded as the origin of low washfastness of the microdenier polyester.

Two methods to reduce the undesirable effect of thermomigration are considered as follows:

(1) The staining ability of disperse dyes towards other fibres in washing liquor needs to be reduced. Although the thermomigrated dyes are dissolved or dispersed in the washing liquor, if the dyes are easily hydrolysed and turned into colourless compounds, they do not cause any problem of staining other fibres at all. The example of this type of dye is benzodifuranone disperse dye, which is easily hydrolysed in weak alkaline aqueous solutions.
(2) Another method to minimise thermomigration is by enhancing the interaction between the fibre and the dye. In absence of any polar groups in polyester fibres, introduction

of bulky substituents or polar groups has proved to be a useful method enhancing the fibre–dye interaction through van der Waals force or hydrogen bonding. It is well known that as the size of the disperse dye becomes big, the washfastness improves. Therefore, the development of disperse dyes with a low degree of thermomigration by increasing molecular size can be considered plausible.

Kim et al. [115] studied the effect of thermomigration on normal and microdenier polyester fibres. They experimented with two different kinds of dyes. It was found that washfastness of the polyester fabrics was closely related to the degree of thermomigration of disperse dye during heatsetting. The degree of thermomigration of the regular size disperse dye such as Dye I (Figure 58) increased with the concentration of dye in the fibre and decreased as the denier of the fibre increased. Thermomigration for bigger sized disperse dye like Dye II (Figure 58) did not change markedly according to the dye concentration and the denier of the fibre. Since the interaction between the polyester fibre and disperse dye were mainly van der Waals forces and hydrogen bonding, it was difficult to inhibit the dye migration completely during heatsetting, but the introduction of two benzene rings into the dye molecule was found to be effective in reducing the degree of thermomigration especially for the microdenier polyester. The authors concluded that the dye having bigger molecular size and higher substantivity was suitable for microdenier polyester to maintain high level of washfastness without a notable diminution in the dye uptake.

Microdenier polyester starts to absorb Dye I at lower temperature (Figure 59) and gets dyed at a rapid rate compared with the conventional polyester. The faster rate might be attributed to the large surface area of the microdenier polyester. The rate of exhaustion of Dye I is very dependent on the thickness of the fibre, but that of Dye II appears to be almost independent on the thickness of the fibre except the early stage of dyeing, suggesting that the increased surface area does not affect markedly the rate of dyeing of the bulky dye that diffuses slowly. It can also be noted that the rate of dyeing of Dye I on 0.5-d fibre is faster than that of Dye II on the same fibre. The results clearly show that the relatively big dye is difficult to penetrate into the fibre.

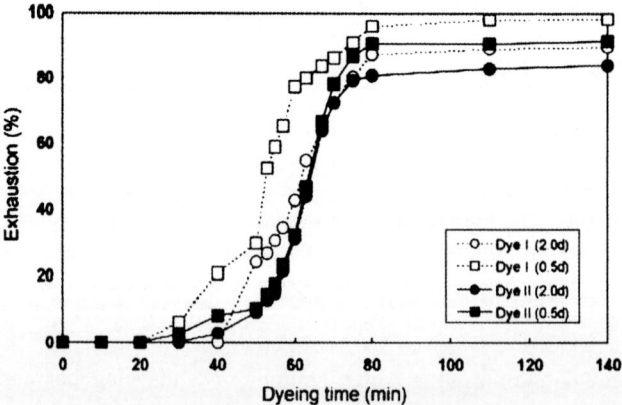

Figure 58. Chemical structure of disperse dyes used in the study.

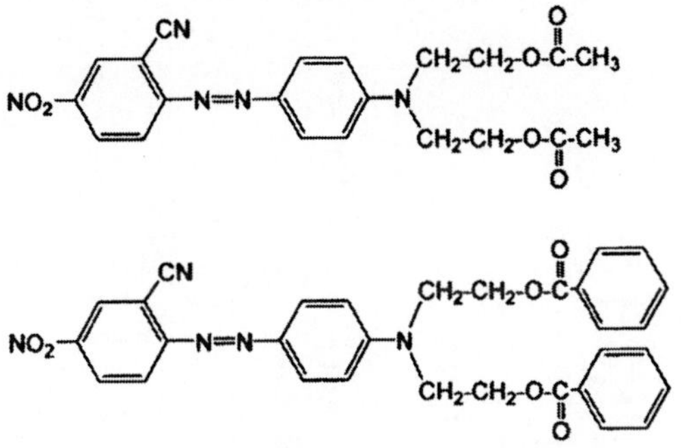

CN
NO₂—〈 〉—N=N—〈 〉—N〈 CH₂-CH₂-O-C-CH₃ (O)
                                CH₂-CH₂-O-C-CH₃ (O)   Dye I

CN
NO₂—〈 〉—N=N—〈 〉—N〈 CH₂-CH₂-O-C—〈 〉 (O)
                                CH₂-CH₂-O-C—〈 〉 (O)   Dye II

Figure 59. The rate of exhaustion curves of disperse dyes applied to conventional and microdenier polyesters at 1% owf (after Kim et al. [115]).

9. Various uses of microfibres

Microfibres have been put to various uses. Two major advantages of microfibres lie in their cleaning and filtration applications. The principle behind these two are discussed below.

9.1 Mechanism of cleaning by microfibres

The surface area of the microfibres is more than 10 times of a normal fibre. The microfibre's small diameter translates into a much larger surface area than that found in conventional fibres. The small diameter of the fibres also provides a particularly powerful capillary action, which, in addition to pulling in liquid, also pulls in particulates and microbes contained within the liquid. Thus, the combination of the increased surface area and capillary action gives the ultra-microfibre cloth the ability to absorb vast amounts of liquid many times its own weight.

The ultra fineness of each fibre also allows more fibres to be packed per square centimetre. This results in a far greater quantity of fibres coming into contact with the surface to be cleaned. This gives a faster and more efficient result. The wedge-shaped microfibres can also trap dirt more easily, as illustrated in Figure 61.

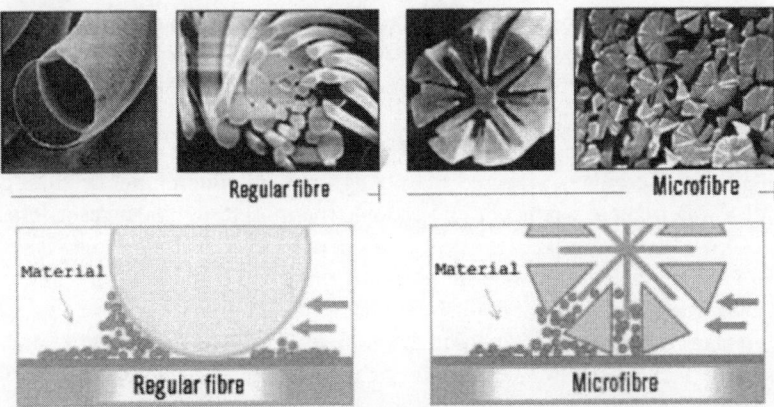

Figure 60. Cleaning properties of a regular fibre, when compared to a microfibre [116].

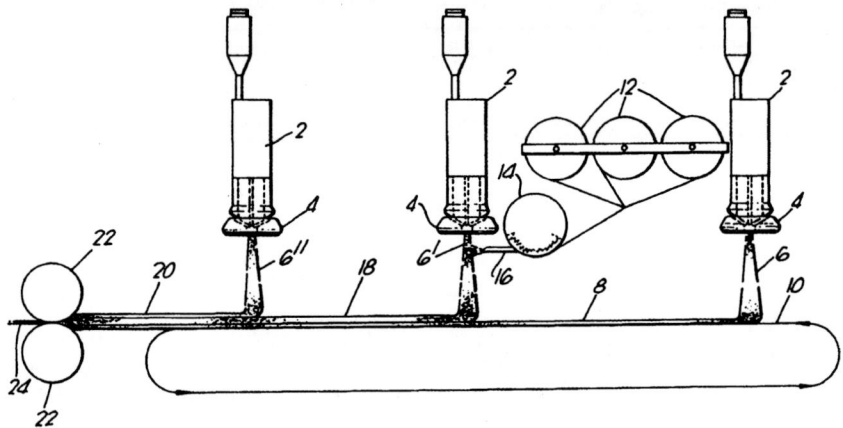

Figure 61. Machinery for manufacturing disposable wipes (after US Patent 4784892).

The cleaning properties of the ultra-microfibres are further enhanced because they have a cationic (positive) charge due to the presence of the polyamide in the ultra-microfibres. Most dirt and dust particles, bacteria, pollen, oxidation on metals, etc., have an anionic (negative) charge. Thus, the ultra-microfibres naturally attract negatively charged particles, bacteria, etc.

9.2 Mechanism of filtration by microfibres

Filtration applications rely on the dominant mechanisms like (1) interception, (2) Brownian motion, (3) coalescense, (4) electrostatic effects, and (5) triboelectric effects [117].

Splittable synthetic microfibres enhance each of these filtration mechanisms and increase the filtration performance of media working in each of these regimes. The very nature of extremely fine diameter fibres makes them suitable for filter applications where interception, Brownian motion, and coalescence are the dominant/significant mechanisms.

The electrostatic effect plays a very important role, and there has been work with filters from microfibres, which have been given a charge. Some splittable fibres spun from two dissimilar polymers give a researcher the choice of polymers and can define a fibre that can be electrostatically charged and/or triboelectrically charged under flow conditions. Thus, these fibres are extremely versatile in their filtration usage and can be the basis for a new series of proprietary filter materials.

In one of the US Patents, Pall et al. [118] reported filter media that had charged resin-coated inorganic microfibres prepared by mixing inorganic microfibres with an aqueous solution of a water-soluble, noncolloidal cationic thermosetting binder resin. The coated microfibres may be used in the dispersed or suspended form as a filter aid. They can be formed into a filter sheet that can be subsequently dried and cured to form a filter sheet of narrowly distributed pore sizes of half a micrometre. In addition, it can have a positive zeta potential in alkaline media. By providing the normally negative zeta potential microfibres with a positive zeta potential the binder resins used to coat the microfibres substantially enhance particulate removal capabilities of the microfibre filter sheet.

The splittable microfibres are more suited as flex-resistant materials. In pulsing applications where the filter medium is continuously flexed but also requires stiffness, splittable synthetic fibres add a high degree of reinforcement to the filter medium because there are at least 16 times the number of fibres available for reinforcement when they are split for segmented fibres.

Some of the most important and interesting uses are underlined below.

9.3 Industrial

A novel product, disclosed in a patent [119], uses microfibres for cleaning up oil spills. The product comprises ultra-fine polymeric fibres that are produced from various polymeric materials by mixing with thermoplastic poly(vinyl alcohol) and extruding the mixture through a die followed by further orientation. The poly(vinyl alcohol) is extracted to yield liberated ultra-fine polymeric fibres. The fibres are ultimately processed into said product, such as a mat, which is placed directly on the oil spill to absorb the oil.

Low-cost wiper material for industrial and other applications having improved water and oil-wiping properties has been reported in a US Patent [120]. A base material of melt-blown synthetic, thermoplastic microfibres is treated with a wetting agent and may be pattern bonded in a configuration to provide strength and abrasion resistance properties while promoting high absorbency for both water and oil. The wiper displays a remarkable and unexpected ability to wipe surfaces clean of both oil and water residues without streaking. It may be produced in a continuous process at a low cost consistent with the convenience of single-use disposability.

A US Patent [121] describes disposable wipes comprising of melt-blown polymeric microfibres intermixed with wood pulp fibres. The wood pulp fibres were interconnected by and held captive within the matrix of microfibres by mechanical entanglement and interconnection of the microfibres and wood pulp fibres. The development of the above idea has been reported in a US patent [122], which describes a nonwoven material useful for disposable wipes comprising a layer of melt-blown polymeric microfibres inter-mixed with fibres of absorbent material and absorbent or super-absorbent particles.

Referring to Figure 61, the first layer of melt-blown microfibres is laid at the right end of the forming wire 10. The central layer is produced partly by the microfibres and partly by wood pulp that is fed from pulp reels 12 to a rotary pulp picker 14, from which the picked pulp is blown through a nozzle 16 to mingle with the melt-blown microfibres.

Such material can readily absorb fluids, including oil, and can subsequently be squeezed out readily. The material also has an integral strength and a substantially lint-free wiping surface.

An air pollution control process [123] employing an improved rotatable collector, which by its position becomes a filtering and adsorbing station and a combustion and desorbing station, and an oxidiser are utilised in an apparatus and process for removing airborne particulate materials and organic vapours from an air stream. The rotatable collector comprises an assembly of alternate layers of refractory microfibres, metal screens, and a thin layer of adsorbent carbon.

Waterproof, multilayered nonwoven fabric of reduced weight having good vapour permeability and the method for its production has been described by Corovin-Gmbh [124]. The fabric comprises at least one layer of coarse, melt-spun, thermoplastic filaments and at least one layer of fine, melt-blown thermoplastic microfibres. The layers are thermally bonded together at intermittent points and, while being heated, are subjected to a force in at least one direction without tearing. The coarse filaments are elongated in the direction of force and the fine microfibres are straightened in the direction of force, in the absence of drawing, to form a denser array of the microfibres having a lesser thickness within the resulting fabric.

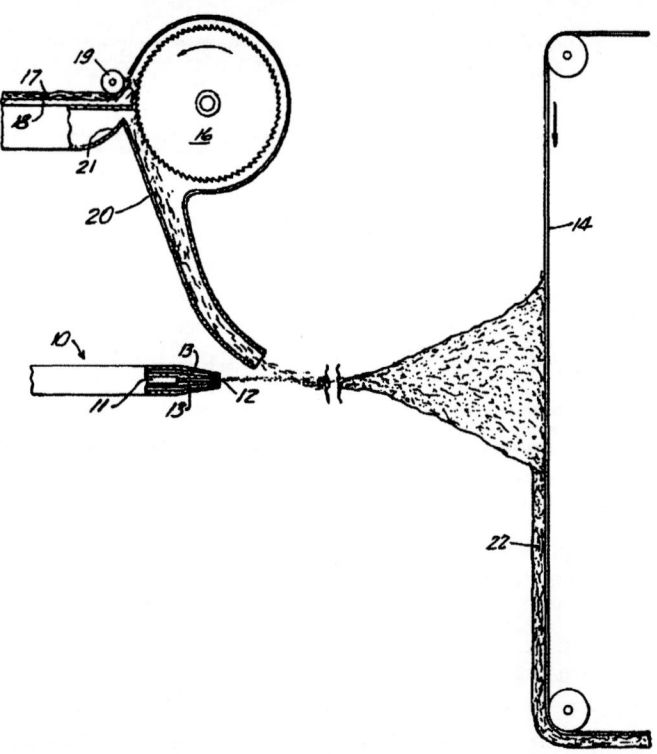

Figure 62. Method for creation of a web of blended microfibres and crimped bulking fibres (after US Pat. 4118531).

A US patent [125] provides a fibrous web, which exhibits high thermal resistance per unit thickness. This web incorporates microfibres (generally less than 10-μm diameter), but only as one component fibre in the web. In addition, a web includes bulking fibres, i.e. crimped, generally larger-diameter fibres, which are randomly and thoroughly intermixed and entangled with the microfibres and account for at least 10% by weight of the fibres in the web. The mechanism of the process, as taken from the above patent, is illustrated in Figure 62.

The crimped bulking fibres function as separators within the web, separating the microfibres to produce a lofty resilient web capable of filling a much larger volume than a conventional microfibre web. This web can be used as an excellent thermal insulator. The researchers of the above patent gave a probable explanation for the high-insulating values of a web of the invention. They proposed that a thin layer of air contacting a fibre or other surface is held by that surface against movement. Since the surface area of microfibres was greater than for larger fibres such as polyester staple fibres, more air is held in place by the microfibres, which results in a reduced transfer of heat within a web containing microfibres. Also, when the microfibres are opened up or spaced apart by the presence of bulking fibres in a web of the invention, the surface area of the microfibres is more effectively used, making it possible to hold more air in place and even further reduce the transfer of heat.

A process and apparatus for manufacturing a microfibre structure for absorbing impact energy is discussed in a patent [126]. The method involves subjecting microfibre threads to a pressurised air jet to open the threads by separating the microfibres into each thread and entangling them to form a mass of loosely entangled microfibres. The pack thus formed is further compacted.

Acoustical insulation materials are used in many applications in various industries, especially automobile where vehicle manufacturers use acoustical insulation in vehicle doors. The technically well-known sound attenuating materials include felts, foams, compressed fibres, glass powder or 'rock wool' and recycled fabrics that have been hammer milled, resinated and thermoset materials. The use of microfibres in acoustical insulation has also been reported in patent literature.

A US Patent provides [127] thermally stabilised polypropylene melt-blown microfibre acoustical insulation web that has a resistance to thermal degradation at a temperature of 135°C for at least 10 days. The polypropylene has a thermal stabiliser uniformly distributed within the melt-blown microfibre polymer which, when produced, is subject to thermal and/or catalytic degradation in the absence of significant levels of thermal stabiliser or antioxidant.

9.4 Civil

Toughness and strength improvements in cement-based matrices due to micro-fibre reinforcement were investigated by Banthia and Sheng [128]. Cement paste and cement mortar matrices were reinforced at 1%, 2% and 3% by volume of different microfibres including polypropylene, and these composites were further characterised in the hardened state under an applied flexural load. Both notched and unnotched specimens were tested in 4-point bending. Considerable strengthening, toughening and stiffening of the host matrix due to microfibre reinforcement were observed. The test data from the notched specimens were used to construct crack growth resistance and crack opening resistance curves for these composites and to identify the conditions necessary for failure. This paper recognised the potential of these composites in various applications and stresses the need for continued research.

An investigation into the reinforcing behaviour of microfibres in cement composites was done by Nelson [129]. Fracture toughness tests were conducted on thin sheet cement composites reinforced with polypropylene (PP), polyvinyl alcohol (PVA), and refined cellulose (RC) fibres under air-dry conditions. Fracture toughness in the absence of microscopic damage was quantified for these composites, as was fracture toughness that included the energy absorbed in the microcracked region before failure crack localisation. These tests revealed that the PVA and RC microfibres were able to effectively postpone microcrack formation, thereby delaying the localisation of the failure crack. The PP microfibres were not able to provide the same level of reinforcement.

The difference in behaviour between the PP and PVA fibres in the frontal process zone was attributed to their interaction with the cement matrix. The PVA fibre has a strong chemical and frictional interfacial bond to the cement matrix, while the PP fibre has only a weak frictional bond [130].

9.5 Medical

Fabrics from microfibres have excellent breathability and have been used for wound care. Ultra-microfibres are generally triangular in cross section, have sharp edges, and have a diameter of approximately 3 μm. A bacterium typically has a diameter of 2–5 μm, so the extremely small size and structure of the ultra-microfibre allow that fibre to get beneath the bacteria or other small microbes and particles that are smaller than the fibre, and substantially remove them from a surface. In addition, to improve performance, the ultra-microfibres are usually mixed with polyester fibres in a 50/50 ratio in the case of woven material, and a 70/30 ratio of polyester to ultra-microfibre in the case of knitted material.

9.5.1 Polypropylene

Polypropylene (PP) microfibre spunbonds have application in wound-care, where they are used as hydrophobic backings to prevent exudates strike-through for extra protection against contamination. At the same time, the air permeability and breathability of these nonwovens promote healing and their softness and flexibility allow excellent adaptation to the skin. In addition, PP microfibre spunbonds have potential application in disposable surgical gowns and masks where spunlaced fabrics are widely used. The barrier properties of these spunbonds are more than 25% better than the spunlaced fabrics at about half their weight (35 g/m^2). Their softness, high permeability and breathability guarantee a high level of comfort in wearing when used as surgical gowns, and for application as surgical face masks; the hydrophobic outer layer prevents fluid strike-through in case of splashes.

9.5.2 Polyethylene

A patent [131] describes a fabric sheet, a curable resin coated onto the fabric sheet and a plurality of microfibre fillers dispersed into the resin. The incorporation of microfibre fillers into the casting materials adds substantially to the strength of the cured casting material, particularly when the fabric used therein is a non-fibreglass fabric.

9.6 Apparel

Evolon, launched by Freudenberg Nonwovens [132], is made from continuous microfibres. The Evolon process has filament spinning and web formation after which high-pressure water jets split the continuous filaments into 0.05–0.2-dtex microfibres. Fabrics intended for garments from Evolon have outstanding drape and handle, are comfortable to wear, can be given hydrophobic or hydrophilic treatments and can be laundered easily.

Kuraray [133], the market leader in the development of synthetic leather, has introduced the new Amaretta JP synthetic suede. The new product satisfies both aesthetic and physical properties criteria. The product is based on a polyester microfibre and a microporous polyurethane resin.

Kanebo Gohsen Ltd. has developed a fabric called Beledano, using Belima SX microfibres. Belima SX is 0.05 denier in comparison with the 0.1 denier of Belima X polyester/nylon ultrafine microfibre used to produce suede-type fabrics. Belima SX is produced by spinning a bicomponent fibre that is then split and divided in a special finishing process. Beledano is currently being marketed for apparel applications, such as coats and jackets, as well as for interiors.

Various microfibre jackets are now in use. For example, a Web site [134] advertises for a 100% polyester sueded microfibre body with water-repellent finish, convertible collar with button-tab closure, full-zip front, slash front pockets with snap closure, locker loop, inside pen pocket, rib knit cuffs and waistband, polyester/cotton lined body. A reversible golf length jacket having 100% polyester sueded microfibre body with water repellent finish is also advertised.

Smith [135] gives a good idea about the garments made from microfibres that are usually labelled to identify their presence, for example: '100% polyester microfibre'. Many fibre companies use trade names to identify their microfibre products. A few examples include the following:

(1) Trevira Finesse (polyester)
(2) Fortrel Microspun (polyester)
(3) DuPont Micromattique (polyester)
(4) Shingosen (polyester)
(5) Supplex Microfibre (nylon)
(6) Tactel Micro (nylon)
(7) Silky Touch (nylon)
(8) Microsupreme (acrylic)

Fabric manufacturers also use trade names for microfibre fabrics. They include the following:

(1) Logantex
    (a) Charisma – dress weight with suede-like finish
    (b) Ultima – water-repellent finish
(2) Thompson of California
    (a) Moonstruck – soft sueded finish, silk-like
    (b) Micromist – brushed finish
    (c) Regal – dry hand
(3) Springs Mills
    (a) Silkmore – sandwashed silk finish
    (b) Stanza – water-repellent microtwill
    (c) Vanessa – reversible fabric for rainwear

9.7 Synthetic leather

Natural leathers, e.g. chamois leather, with beautiful appearance, softness, and a porous structure that give them high water absorption and vapour permeability, are popular in the market. However, these are less and less in markets because of their restricted source, exorbitant price, as well as more and more awareness of protecting animals. As an alternative, synthetic leather based on nonwoven support material coated with polyvinylchlorides or polyurethanes is usually used for habiliments and shoe upper materials [136].

In a typical processing procedure for synthetic leather, the nonwoven fabric of the sea–island microfibre with a reticulate structure is treated by wet-cured polyurethane process, followed by extraction of the matrix polymer using an organic solvent. Some properties of the synthetic leather are even better than those of natural leathers, such as soft handle, crease resistance, strength and elongation at break, hygroscopic ability, ventilation and easy-care [137].

The sea island–type microfibres have become popular amongst all the varieties. In a work by Chen et al. [138], a gradient extraction method was developed and used for synthetic leather made of polyurethane (PU) and microfibre polycaprolactam (PA-6), by which the two components were separated using DMF and formic acid as solvents, respectively. Figure 64 shows the cross section of the synthetic leather developed.

Some properties of the synthetic leather are even better than those of natural leathers, such as soft handle, crease resistance, strength and elongation at break, hygroscopic ability, ventilation, easy-care, easy volume-producing, etc. [139]. Synthetic leather, however, has poor dye fastness resulting from larger surface area of micro-fibre, when compared with

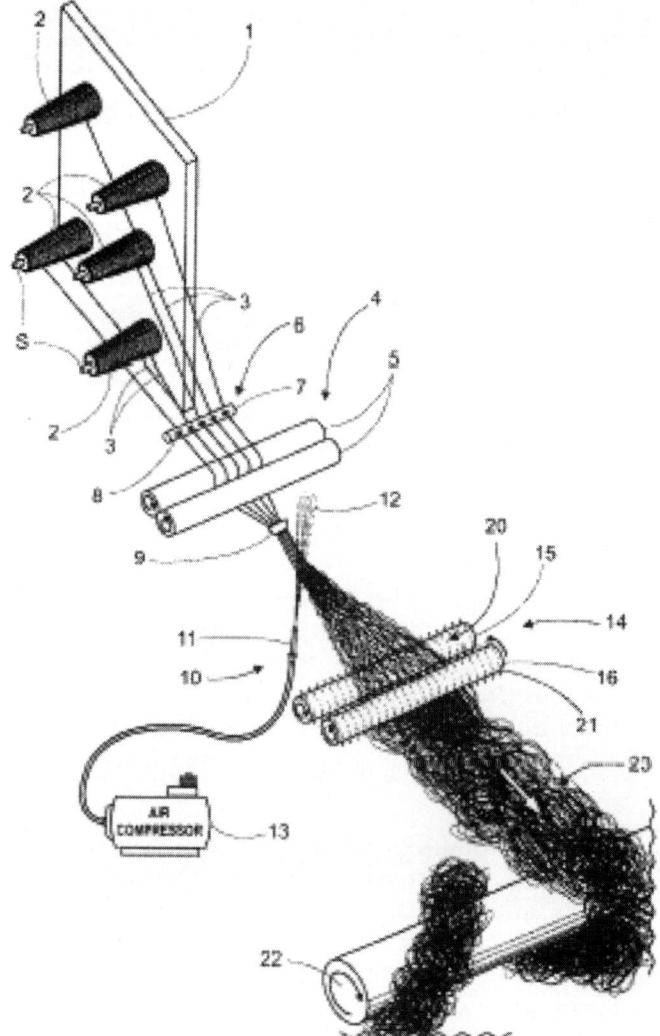

Figure 63. Microfibres for absorbing impact energy (after US Pat. 6684468).

those of the natural ones and other reasons that have been discussed in detail in the 'Dyeing' section.

9.8 Household

A super-absorbent drying towel [140] uses microfibre that eliminates the need to use any soap or detergent as claimed by the producer. The microfibres with their enhanced surface area grab the grease and dirt. It also gives much better cleaning results in less time.

Another company [141] advertises for super-microfibre fabrics that consist of 0.2–0.3 denier synthetic yarns such as polyester and N/P mixture. The surface of the microscopic fibres gently cleanses pores without irritating delicate skin. The water retention capacity is 2–3 times more than that of normal cotton fabrics. The fabric has been successfully used for wiping clothes for cleaning kitchen & utensils/household

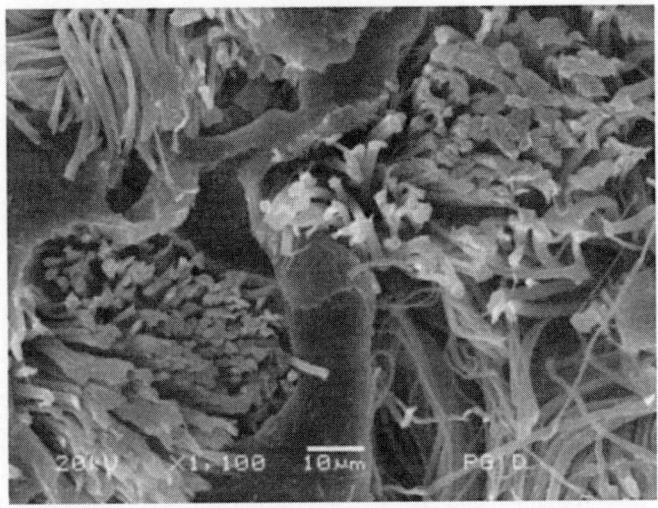

Figure 64. Synthetic leather.

furnishings/cars/mirrors/windows, cleaning of eye glasses, as hair towels/bath towels/face cloths/exfoliating towels/sports towels/mittens.

9.9 Miscellaneous

Microfibres have limited tendency to lint. Microfibre towels last much longer than conventional towels. Microfibre is also extremely absorbent. One of the companies claim that drying towels [142] and Ultimate Microfibre Chamois' can absorb 10 times their weight in water.

Improved absorbent products are reported in a patent [143] based on microfibre technology. The absorbent products comprise pressure-sensitive microfibres that provide good liquid transport properties and resiliency and mask the odours associated with bodily fluids.

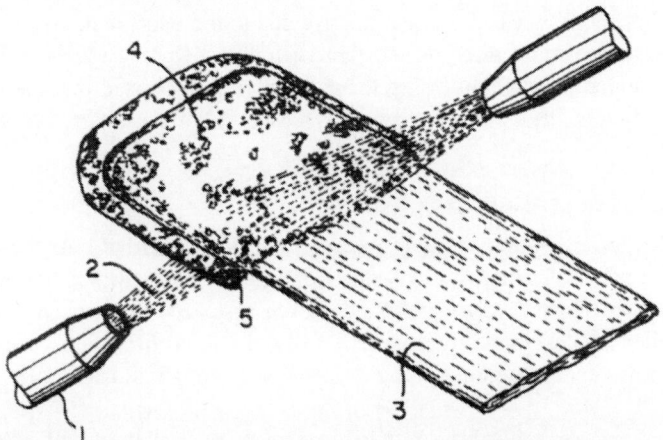

Figure 65. Absorbent structure using microfibres (after US Pat. 5,885,681).

In the figure, from US Patent [144], the absorbent core '3' was coated with microfibres '2' on the surface by using a spray nozzle '1' to form a liquid permeable facing. The spraying techniques can use the spray nozzle, which results in spiralling fibres or a melt blowing technique which straightaway deposit the microfibres on the surface.

The microfibres of less than 1 dtex are formed using molten spray technology and are used to form coatings on the substrate of 0.002–0.084 g/m^2. They are sprayed onto the absorbent core, using a spray nozzle or a melt-blown technique. They can be used in sanitary napkins, incontinence garments, and disposable diapers using less expensive techniques.

Sterling Fibres [145], along with Glenoit Corporation, developed microfibre pile, utilising MicroSupreme$^{®}$ high-tech acrylic. In construction of pile fabrics, a sliver is created and then knit into a backing and is then eventually sheared to the specific pile height desired. Jacquard designs may be knit into the fabric, resulting in a very precise, crisp, clean and long lasting pattern. These fabrics are known today as MicroBerber$^{®}$, MicroGlenaura$^{®}$, MicroFabric$^{®}$, MicroLana$^{®}$ and GlenPile$^{®}$. These micro pile fabrics are used in high-tech and fashion outdoor lines, branded coat, sportswear, accessories (hats, gloves, scarves), footwear, robes and loungewear, as well as in home textiles and upholstery. New lighter weight fabrics are ending up in spring/summer applications such as golf.

A patent [146] describes composite sheets made of ultrafine sheath-core composite fibres. The composite fibres have a fineness in the range of 0.0001–0.5 denier and a core/sheath weight ratio in the range of 10/90–70/30. When the composite fibres are combined with an elastic material such as polyurethane, suede-like artificial leathers having excellent softness, touch, feel and colour can be obtained.

10. Economics of Microfibre Processing

The price of microfibre is generally 5%–10% higher than conventional fibres that will push up the yarn price by 4%. However, the conversion of fibre to yarn will come down due to increased production per spindle at ring frame and fly frame owing to reduction in twist by 5%–10%. Also, due to better working of ring frame, the work assignment of the ring frame tenters can be increased to some extent. The final net increase in yarn cost will come down to 2%. As against the 2% increase in yarn cost, there are many advantages. The products made out of microfibre yarns are more attractive due to their soft feel, greater absorbency, etc. The fabrics fetch more price due to their specially nature. Besides it is possible to produce value added products by using these fibres. It can be said that the use of micro fibres is economically attractive if the products have good marketability [147].

11. Limitations and precautions

The most important limitation of microfibres is their heat sensitivity. As the fibre strands are so fine, heat penetrates more quickly than with thicker conventional fibres. As a result, the microfibres are more heat sensitive and scorch or glaze if too much heat is applied or if it is applied for too long a period. Generally, the microfibres are wrinkle resistant, but if pressing is needed at home or by drycleaners, care should be taken to use lower temperatures.

Microfibres can generally be cared for in a manner similar to that of conventional fibres made from the same fibre type. For example, fabrics made from polyester and nylon microfibres can probably be machine washed and tumble dried in a manner similar to fabrics made from regular polyester and nylon fibres. Polynosic or high-wet modulus rayons are machine washable while viscose rayons perform best when dry cleaned. Rayon microfibre should be cared for depending on whether it is a polynosic or viscose-type rayon. The fibre properties, not the fineness of the fibre, usually dictate recommended care. Always follow care labels on garments.

A few cautions should be noted regarding microfibres. As they are very fine or of small diameter, heat penetrates the fibres more quickly than thicker fibres. As a result, glazing, melting or scorching can occur quickly. This is a particular concern with heat-sensitive fibres such as polyester or nylon. It is preferable to use a cool iron, if pressing is necessary, and the iron should not be left on the fabric for too long. It is also advisable not to apply too much pressure, which can create shine and ridges on the surface.

Static may develop in fabrics from synthetic microfibres, especially during dry winter months when heating systems are turned on and the humidity is low. Fabric softeners in the rinse cycle of the washing machine may lessen the problem. Paper dryer sheets can be used; however, temporary spots from excessive heat in the dryer may form on the microfibre. The delicate finish of microfibre fabrics and the amount of fibre surface make the spots noticeable if they develop.

Static may develop in fabrics from synthetic microfibres, especially during dry winter when heating systems are turned on and the humidity is low. Fabric softeners in the rinse cycle of the washing machine may lessen the problem. As with all fine garments, rough or jagged jewellery should be avoided. It can cause pulls, snags or general abrasion to garments. Although microfibres in a yarn are strong, the individual fibres are extremely fine and could abrade easily.

12. Conclusion and suggestions for further work

Since the mid-1980s the market importance of microfibres has been rising throughout the world. They are opening up new marketing opportunities to the fabric manufacturers and inspire fashion designers to new creations. Microfibre holds a lot of promise for the future. The much-friendlier properties of microfibres as compared to those of other synthetic fibres have been able to revitalise the synthetic fibre market.

There are several areas that require further attention. From the viewpoint of a fibre and polymer scientist, various microfibres are yet to be fully characterised. Studies aiming at the structural characterisations due to stress and treatment temperature are necessary. The understanding of the macromolecular characteristics of the microfibre would give important inputs for the downstream processing machinery. There have been some research on use of microfibres in the medical sector but there is still a lot to be done. The high surface area can be a real boon for various medical components that rely solely on surface activity.

Weaving of microfibres is a field that calls for more attention. Weaving such yarns in shuttleless looms in high speed is crucial without breaks. This demands a much better yarn-processing machinery. Splicing of yarns from microfibres in high-speed autowinders needs more attention and change of settings in splicers like air pressure, time of spray of air, demands more intensive studies for productive operations. The dyeing behaviour of the microfibres is a critical field and needs additional systematic studies.

Microfibres have been put to various uses but still many areas remain to be explored. There have been complaints of static generation from furnitures using fabric from microfibre. More research needs to be directed at the production of durable anti-static finish on such fabrics. Use of cellulosic microfibres as reinforcement with biodegradable matrices is one such important and potential area that needs attention. Extensive research is also called for to produce these fibres at a much lower cost.

References

[1] A. Kasei, Teijin, Unitika, Kuraray, and Mitsubishi Rayon, Toyobo, *private communications*, June 1991.

[2] K. Fujimoto, K. Iohara, S. Owaki, and Y. Murase, *The change of end groups of polyester fibres by caustic treatment*, Scn-i Gakkaishi 44 (1988), pp. 477–480.

[3] A. Suzuki and N. Mochizuki, *Mechanical properties and microstructure of poly(ethylene terephthalate) microfiber prepared by carbon dioxide laser heating*, J. Appl. Polym. Sci. 90 (2003), pp. 1955–1958.

[4] A. Suzuki and S. Narusue, *Isotactic polypropylene microfiber prepared by continuous laser-thinning method*, J. Appl. Polym. Sci. 92 (2004), pp. 1534–1539.

[5] A. Suzuki and K. Kamata, *Nylon 6 microfiber obtained by a continuous-thinning method with a carbon dioxide laser*, J. Appl. Polym. Sci. 92 (2004), pp. 1449–1453.

[6] A. Suzuki and T. Hasegawa. *High temperature zone-drawing of nylon 66 microfiber prepared bv CO_2 laser-thinning*, J. Appl. Polym. Sci. 99 (2006), pp. 802–807.

[7] A. Suzuki and S. Narusue, *Isotactic polypropylene microfiber prepared by carbon dioxide laser-heating*, J. Appl. Polym. Sci. 99 (2006), pp. 27–31.

[8] A. Suzuki and M. Tojyo, *Poly(ethylene-2,6-naphthalate) microfiber prepared by carbon dioxide laser-thinning method*, Eur Polym. J. 43 (2007), pp. 2922–2927.

[9] A. Suzuki and H. Ohnishi, *Isotactic polypropylene hollow microfibers prepared by CO_2 laser-thinning*, J. Appl. Polym. Sci. 102 (2006), pp. 2600–2607.

[10] A. Suzuki and M. Kishi, *Preparation of poly(ethylene terephthalate) nonwoven fabric from endless microfibers obtained by CO_2 laser-thinning method*, Polymer 48 (2007), pp. 2729–2736.

[11] A. Suzuki and D. Mizuochi, *Zone-drawing and zone-annealing of poly(L-lactic acid) microfiber prepared by CO_2 laser-thinning method*, J. Appl. Polym. Sci. 102 (2006), pp. 472–478.

[12] F. Die'val, D. Mathieu, and B. Durand, *Comparison and characterization of polyester fiber and microfiber structure by X-ray diffractometry and viscoelasticimetry*, J. Text. Inst. 95 (2004), pp. 131–146.

[13] www.hillsinc.net (accessed 5 January 2008).

[14] www.hillsinc.net/nanofibre.shtml (accessed 5 January 2008).

[15] Gillespie, B. Don. Private Communication. Fleissner Incorporated, 12301 Moores Chapel Road, Charlotte, NC 28214.

[16] http://www.premiumautocare.com/microfibre-products.html (accessed 10 January 2008).

[17] Pike, Richard Daniel, Sasse, Philip Anthony, White, Edward Jason, Stokes, and Ty Jackson, US Pat. No. 5759926 (to Kimberly-Clark Worldwide, Inc., Neenah, WI), 2 June 1998.

[18] Pike, U.S. Pat. No. 6,624,100 (to Kimberly-Clark Worldwide, Inc., Neenah, WI), 23 September 2003.

[19] www.fitfibres.com/files/Microfibres%20for%20Filtration.doc (accessed 10 January 2008).

[20] C. Sun, D. Zhang, Y. Liu, and J. Xiao, J. Ind Text. 34, (2004), 17.

[21] H. John, *Trends in barrier fabrics for medical applications*, Int. Fibre J. 16(4), (2001) pp. 50–51.

[22] H. Lee, H. Seong, J. Kwang, Y. Dae, and Y. Han, Text. Res. J. 74(3) (2004), pp. 271–278.

[23] V.L. Gibson and R. Postle, *An analysis of the bending and shear properties of woven, double-knitted, and warp-knitted outerwear fabrics*, Text. Res. J. 48(1) (1978), pp. 14–27.

[24] C. Bryan, W. Ya, R. Hurt, and T. Chattanooga (to E I Du pont de Nemours and Company, Wilmington, Del), US Pat. No. 3381074, April 1968.

[25] M. Okamoto, K. Watanabe, Y. Nukushina, and T. Aizawa, US Pat. No. 3,531,368 (to Toyo Rayon Inc), 29 September 1970.

[26] M. Okamoto, K. Ashida, K. Watanabe, and S. Taniguchi, US Pat. No. 3,692,423 (to Toray Industries, Inc., Tokyo, Japan, 19 September 1972.

[27] Y.O. Moriki and M. Ogasawara, Japan, US Pat. No. 4,445,833 (to Toray Industries, Inc., Tokyo, Japan), 1 May 1984.

[28] T. Nakajima, *Advanced Fibre Spinning Technology*, Woodhead Publishing Limited, Cambridge, England, 1994, 199.

[29] H.Y. Park, L.J. Woo, M.H. Kim, US Pat. No. 4460649 (to Kolon Industries, Inc., Seoul, Republic of Korea), 17 July 1984.

[30] T. Nakajima, *Advanced Fibre Spinning Technology*, Woodhead Publishing Limited, Cambridge, England, 1994.

[31] L.C. Wadsworth and A.O. Muschelewicz, *Book of Papers*, Fourth International Conference on Polypropylene Fibres and Textiles, Nottingham, England, September 1987, pp. 47.11–47.20.

[32] B.D. Haynes, *An Experimental and Analytical Investigation on the Production of Microfibres Using a Single Hole Melt Blowing Process*, PhD Dissertation, The University of Tennessee at Knoxville, 1991.

[33] H. Yin, Z. Yan, and R.R. Bresee, *Experimental study of the meltblowing process*, Int. Nonwovens J. S(l) (1999), pp. 60–65.

[34] M.A.J. Uyttendaele and R.L. Shambaugh, *Polymer meltblowing*, AlChE J. S6(2) (1990), p. 175.

[35] M.W. Milligan and B.D. Haynes, *The use of crossflow to improve nonwoven meltblown fibers*, J. App. Polym. Sci. 58 (1995), pp. 159–163.

[36] G. Najour and G. Ward, US Pat. No. 6379136, 30 April 30 2002.

[37] http://cerig.efpg.inpg.fr/tutoriel/non-tisse/photo22.htm (accessed 10 January 2008).

[38] www.fibre2fashion.com (accessed 10 January 2008).

[39] US Pat. No. 3,497,918, S. Pollock, Richmond Va, G. Smith (to Du Pont), 3 March 1970.

[40] US Pat. No. 3,549,453, G. Smith (to Du Pont), 22 December 1970.

[41] US Pat. No. 3, 593,074, L. Isakoff (to Du Pont), July 1971.

[42] V.E. Kalayci, P.K. Patra, A. Buer, S.C. Ugbolue, Y.K. Kim, and S. Warner, *Fundamental investigations on electrospun fibers*, J. Adv. Mater. 36 (2004), pp. 43–47.

[43] A.L. Yarin, S. Koombhongse, and D.H. Reneker, *Taylor cone and jetting from liquid droplets in electrospinning of nanofibers*, J. Appl. Phys. 90 (2001), pp. 4836–4846.

[44] D.H. Reneker, A.L. Yarin, H. Fong, and S. Koombhongse, *Bending instability of electrically charged liquid jets of polymer solutions in electrospinning*, J. Appl. Phys. 87 (2000), pp. 4531–4547.

[45] L. Larrondo and R.S.J. Manley, *Electrostatic fiber spinning from polymer melts. I. Experimental observations on fiber formation and propertie*, J. Polym. Sci., Part B: Polym. Phys. 19 (1981), pp. 909–920.

[46] L. Larrondo and R.S.J. Manley, *Electrostatic fiber spinning from polymer melts. II. Examination of the flow field in an electrically driven jet*, J. Polym. Sci., Part B: Polym. Phys. 19 (1981), pp. 921–932.

[47] L. Larrondo and R.S.J. Manley, *Electrostatic fiber spinning from polymer melts. III. Electrostatic deformation of a pendant drop of polymer melt*, J. Polym. Sci., Part B: Polym. Phys. 19 (1981), pp. 933–940.

[48] R.S. Givens, H. Kenncorwin, G.F. J. Rabolt, and D.B. Chase, *High-temperature electrospinning of polyethylene microfibers from solution*, Macromolecules 40 (2007), pp. 608–610.

[49] P.Q. Pham, U. Sharma, and G. Antonios, *Mikos electrospun Poly(ε-caprolactone) microfiber and multilayer nanofiber/microfiber scaffolds: Characterization of scaffolds and measurement of cellular infiltration*, Biomacromolecules 7 (2006), pp. 2796–2805.

[50] M. Okamoto, *On ultra fine fibres*, Sen-i Gakkaishi, 32 (1976), p. 318.

[51] M. Okamoto, *On ultra-fine fibre and its application*, JTN, p. 94, November 1977; JTN, January 1978, p. 77.

[52] M. Okamoto, *Progress of artificial leather*, Gendai Kagaku, August 1981, p. 52.

[53] M. Okamoto, Reprints from *Japan–China Bilateral Symposium on Polymer Science and Technology*, 1981, p. 256.

[54] M. Okamoto, Chemifasern Textilindustrie, 29/81, 30 and E4, 1979.

[55] M. Okamoto, Chemifasern Textilindustrie, 29/81, 175 and E25, 1979.

[56] http://ohioline.ag.ohio-st (accessed 10 January 2008).

[57] R. Tedesco, L. Console, and M. Cavallini, 35th International MMF Congress Proc. Conference, Dornbirn, September 1996.

[58] F.R. Cook, C.M. Cunningham, US Pat. No. 4753843: Kimberly-Clark Corporation, Neenah, WI, 28 June 1988.

[59] D.A. Early, W.D. Cawlfield, T. Walker, T.P. Weisman, US Pat. No. 4468428 (to The Procter & Gamble Company, Cincinnati), 28 August 1984.

[60] T. Toledano-Thompson, M.I. Loría-Bastarrachea, and M.J. Aguilar-Vega. Carbohydr. Polym. 62(1), 17 October 2005, pp. 67–73.

[61] D. Bhattacharya, L.T. Germinario, and W.T. Winter, *Isolation, preparation and characterization of cellulose microfibres obtained from bagasse*, Carbohydr Polym (2007), doi: 10.1016/j.carbpol.2007.12.005.

[62] F.C. Helms, Jr., M.O. Ilg, R.D. Kent, B.M. Hoyt, and A.J. Hodan, US Pat. No. 5948528 (to BASF Corporation, Mt. Olive, NJ), 7 September 1999.
[63] W. Stibal, 35th International MMF Congress Proc Conference, Dornbirn, 1996.
[64] W. Peschke, D. Schilo, and M. Weber, Chem. Fibres Int. 62 (1997), pp. 197–218.
[65] M.H. Behery, *Effect of Mechanical and Physical Properties on Fabric Hand*, The Textile Institute, UK, Woodhead Publishing Limited, 2005.
[66] F. Leifield, *Carding of microfibres*, Asian Text. J. November (1992), pp. 42–50.
[67] J. Hwang, S.A.O. Wand, *Carding of microfibres*, J. Text. Apparel Tech. Manag. 1(2) (2001), pp. 1–9.
[68] A. Schenek and H. Schwippl, *Processing properties of polyester micro fibres in COM4 – staple fibre spinning*, Milliand Int. 11(28–30), (2005) pp. 42–50.
[69] F. Leifield., *Carding of microfibres*, Asian Text. J. November 1992.
[70] Hainz Ernst and Rotor Yarns, *A look into the future*, Melliand Textilber, 74(4) (1993), E117–E120.
[71] Rieter.LINK 48.2/2006 Tencel Microfibers—Properties and Processing.
[72] Murata Machinery Ltd., *Murata: Spinning microfibre yarns on the MJS system*, Text World, 4 (1994), pp. 42–48.
[73] Y.A. Korkmaz and M.H. Behery, *Drafting dynamics of fine denier polyester fibres*, Text. Res. J. 74(6) 2004, pp. 497–501.
[74] D.S. Taylor, *Some observations on the movement of fibres during drafting*, J. Text. Inst. 45 (1954), T310–T322.
[75] J. McVittie and A.E. Barr, *Fibre motion in roller and apron drafting*, J. Text. Inst., 52 (1961), pp. T147–T156.
[76] Y. Korkmaz, *The effect of fine denier polyester fibre fineness on dynamic cohesion force*, Fibres Text. 12(1) (2004), p. 45.
[77] A. Basu et al., *Study of quality of man-made fibre yarns and blended yarns*, Indian Text. J., May 2005, pp. 21–31.
[78] GertBock, *Open-end rotor spinning of polyester microfibres*, Int. Text. Bull. Yarn Fabric Forming, February 1993.
[79] II. Ernst and R. Yarns, *A look into the future*, Melliand Textilber, 74(4) (1993), E117–E120.
[80] R. Rajamanickam et al., *Interaction of process and material parameters in Air jet spinning*, Text. Res. J., 68 (1998), pp. 708–714.
[81] Rajamanickam et al., *Effect of material and process parameters on the properties of micro denier polyester/cotton blended yarns*, J. Text. Inst. 89 (1998), pp. 243–265.
[82] M. Nikolie et al. *Compact spinning for improved quality of ring spun yarns*, Fibres Text., 11(4) (2003), p. 43.
[83] A. Schenek and H. Schwippl. *Processing properties of polyester Micro fibres in COM4 – Staple fibre spinning*, Milliand Int. 11 (2005), pp. 28–30.
[84] Y. Huh, Y.R. Kim, and W. Oxenham, *Analyzing structural and physical properties of ring, rotor and friction spun yarn*, Text. Res. J. 72 (2002), pp. 156–163.
[85] G. Ramakrishnan, J. Srinivasan, S. Ganesan, and D. Sudha, *A Study on Fibre Migration in Ring and Open-end Micromodal Yarns*, Project Thesis, submitted to Anna university, Kumaraguru college of Technology, 2006. Unpublished work
[86] A. Basu, *Microfibres – Properties, processing and uses*, Mill control report No. 17.
[87] G. Voswinckel, *Sizing of micro filament yarns*, Asian Text. J., December (1992), pp. 60–62.
[88] G.G. Nau, *Requirements of weaving of microfibre yarns*, Int. Text. Bull. Yarn Fabric Forming, February 1994.
[89] G. Ramakrishnan, B. Dhurai, and S. Mukhopadhyay, *An investigation in to the properties of knitted fabrics made from viscose microfibres*. J. Apparel Technol. Manag. In press.
[90] A. Karolia and N. Paradkar, *Properties of knitted Microfibre Fabrics, Part 1: Growth and elastic recovery properties*, J. Text. Assoc., May–June (2004), pp. 31–34.
[91] R. Matic-Leigh, *Dimensional stability, asthetic and mechanical properties of microfibre blended knitted fabrics*, National Textde Center Annual Report, 30 September 1993.
[92] G. Ramakrishnan and S. Mukhopadhyay, *Study on knitted fabrics from polyester microdenier fabrics*, J. Text. Inst. 97 (2007), pp. 31–35.
[93] Sportswool Pro™ The Wookmark Company, www.wool.com (accessed 10 January 2008).
[94] G. Ramakrishnan and Nandhini, *Design and development of microfibre socks*, Project Report, Kumaraguru College of Technology, India (unpublished work).

[95] A.D. Sule, M.K. Bardhan, and R.K. Sarkar, *Development of sports wear for Indian conditions*, Manmade Text. India (2004) p. 123–128.

[96] C.O. Hauer, *Micro fibres in knitted fabrics*, Knitting Tech. 12 (1990), pp. 288–289.

[97] Y.-L. Hsieh, A. Miller, and J. Thompson, *Wetting pore structure, and liquid retention of hydrolyzed polyester fabrics*, Text. R. J. 66 (1996), pp.1–10.

[98] Y.-L. Hsieh and L. Cram, *Enzymatic hydrolysis to improve wetting and absorbency of polyester fabrics*, Text. Res. J. 68(5) (1998), pp. 311–319.

[99] S.H. Kim, S. Kim, and K. Oh, *Water absorption and mechanical properties of pile-knit fabrics based on conjugate N/P microfibers*, Text Res. J. 73(6) (2003), pp. 489–495.

[100] M. Huh and J. Yoon, *The decomposition kinetics of polyester microfibre fabrics by sodium glycerolate/glycerol solution*, J. Appl. Polym. Sci. 64(6) (1998), pp. 1217–1223.

[101] T. Nakamura, S. Ohwaki, and T. Shibusawa, *Dyeing properties of polyester microfibers*, Text. Res. J. 65(2) (1995), pp. 113–118.

[102] S.M. Burkinshaw, K.D. Maseka, and S.D. Cox, *The dyeing of nylon 6.6 microfibre*, Dyes Pigments 30 (1996), p. 105.

[103] J. Koh, M.-J. Oh, H. Kim, and S.D. Kim, *Alkaline dissolution monitoring of radial-type polyester microfiber fabrics by a cationic dye-staining method*, J. Appl. Polym. Sci. 99 (2006), pp. 279–285.

[104] M.J. Drews and C. Jordan, *Kinetics of dyeing with polyester*, Text. Chem. Colourist 30 (1998), p. 13.

[105] T. Nakamura, R.R. Bommu, and Y. Kamiishi, *Dyeing properties of a polyester ultrafine fiber*, Text. Res. J. 70 (2000), pp. 961–964.

[106] F. Dieval, D. Mathieu, P. Viallier, and B. Durand, *Polyester fiber and microfibre structure by X-rays and viscoelasticimetry*, Text. Res. J. 71 (2001), p. 239.

[107] K. Chen, Z. Chen, and J. Xing, *Analyzing the dyeing behavior and chromaticity characteristics of polyester microfibers*, Text. Res. J. 72 (2002), pp. 367–370.

[108] H. Kobsa, B. Rubin, and S. Shearer, *Using optical ray tracing to explain the reduced dye yield of microdenier yarns*, Text. Res. J. 63(8) (1993), pp. 475–479.

[109] H. Kobsa, B. Rubin, S.M. Shearer, and E.M. Schulz, *Using optical ray tracing to explain the reduced dye yield of microdenier yarns*, Text. Res. J. 63 (1993), pp. 475–479.

[110] D.L. Filkin, H. Kobsa, B. Rubin, and S.M. Shearer, *Method for determining and controlling fibre luster properties*, US Patent Application Serial No. 07/526, 853, 1990.

[111] R. McGregor, R.H. Peters, and K. Varol, *The physicochemical hydrodynamics of dyeing*, J. Soc. Dyers Colour. 86 (1970), pp. 437–445.

[112] T. Shibusawa, T. Endo, Y. Kameta, and P. Rys, *Estimation of the thickness of diffusional boundary layer by analyzing rate of dyeing data and by means of multiple membrane layer method*, Sen'i Gakkaishi 42 (1986), pp. T671–T-679.

[113] H. Kobsa, B. Rubin, S.M. Shearer, and E.M. Schulz, *Using optical ray tracing to explain the reduced dye yield of microdenier yarns*, Text. Res. J. 63 (1993), pp. 475–479.

[114] P. Leadbetter and S. Dervan, *The microfibre step change*, J. Soc. Dyers Colour. 108 (1992), pp. 369–371.

[115] S. Kim, M. Kim, B. Lee, and K. Lee, Fibres Polym. 5(1) (2004), pp. 39–43.

[116] http://www.norwexforyourhealth.ca/index.php?pr = Microfibre_Use_&_Care (accessed 10 January 2008).

[117] C.E. Homonoff. *Utilizing multilayered materials for filtration and separation*, Filtration News, March/April 2000.

[118] D. Pall et al., US Pat. No. 4734208, 29 March 1988.

[119] M.L. Robeson, J.R. Axelrod, and A.T. Manuel. US Pat. No. 5120598 (Air Products and Chemicals, Inc., Allentown, PA) 9 January 1992.

[120] H.G. Meitner, US Pat. No. 4307143 (to Kimberly-Clark Corporation, Neenah, WI), 22 December 1981.

[121] R. Anderson, R. Sokolowski, and K. Ostermeier, US Pat. No. 4,100,324 (to Kimberley Clark Corporation), 11 July 1978.

[122] G.D. Storey and P. Maddern, US Pat. No. 4784892 (to Kimberly-Clark Corporation, Neenah, WI), 15 November 1988.

[123] D.G. Foss, River Falls, WI, US Pat. No. 4415342 (to Minnesota Mining and Manufacturing Company, St. Paul, MN).

[124] E. Corovin-GmbH, Official Gazette of the US Patent and Trademark Office Patents, 1999, 1218.

[125] R.E. Hauser, US Pat. No. 4118531 (to Minnesota Mining and Manufacturing Company, St. Paul, MN), July 1978.
[126] Lujin, US Pat. No. 6684468, 3 February 2004.
[127] D.M. Swan, A.R. Ebbens, Patent number: 5961904, Issue date: 5 October 5 1999, Minnesota Mining and Manufacturing Co.
[128] N. Banthia and J. Sheng, *Fracture toughness of micro-fibre reinforced cement composites*, Cement & Concrete Composites (UK), 18(4) (1996), pp. 251–269.
[129] K. Nelson, *Fracture toughness of microfibre reinforced cement composites*, J. Mat. Civil. Eng. 14(5) (2002), pp. 384–391.
[130] T. Kanda and V.C. Li, *Interface property and apparent strength of high strength hydrophilic fibre in cement matrix.*' J. Mater. Civil. Eng., 10(1) (1998), pp. 5–13.
[131] T.M. Scholz, US Pat. No. 5474522 (to Minnesota Mining and Manufacturing Company, St. Paul, MN), December 1995.
[132] High Perform. Text. 12 (2000).
[133] N. Tanaka and J. Tanaka, *Textile systems and Breathable films*, Melliand Int. 6 (2000), pp. 137–141.
[134] http://www.vcctv.com/jacket.htm (accessed 10 January 2008).
[135] A. Joyce Smith., Ohio State University Fact Sheet, http://ohioline.osu.edu/hyg-fact/5000/5546.html (accessed 10 January 2008).
[136] J. Hemmrich, J. Fikkert, and M. van den Berg, *PVC coated support material*, J. Coated Fabrics, 22 (1993), pp. 268–270.
[137] G. Murlasits and G.J. Wlasitsch, *Natural leathers*, J. Coated Fabrics 14 (1985), pp. 172–176.
[138] M. Chen, D.-L. Zhou, Y. Chen, and P.-X. Zhu, *Analyses of structures for a synthetic leather made of polyurethane and microfiber*, J. Appl. Polym. Sci. 103 (2007), pp. 903–908.
[139] G. Murlasits and G.J. Wlasitsch, *Natural leathers*, J. Coated Fabrics 14 (1985), pp. 172–176.
[140] http://www.kitchen-classics.com/mysticma.htm (accessed 10 January 2008).
[141] http://www.fcc.co.jp/tomen-ht/super.html (accessed 10 January 2008).
[142] http://www.ultimatemicrofibre.com/about.cfm (accessed 10 January 2008).
[143] R. Korpman, Bridgewater, US Pat. No. 5885681 (to MCNEIL-PPC, INC), 23 March 1999.
[144] R. Korpman, US Pat. No. 5,885,681, 23 March 1999.
[145] http://www.sterlingfibres.com/pile.htm (accessed 10 January 2008).
[146] O. Miyoshi, I. Hiromichi, and M. Akito, US Pat. No. 4557972 (to Toray Industries, Inc., Tokyo, Japan), 10 December 1985.
[147] A. Basu, *Microfibres—Properties, Processing and uses*, Mill control report No. 17.
[148] A.K. Teijin, K. Unitika, and M.R. Toyobo, *private communications,* June 1991.
[149] T. Nakajima, *Advanced Fibre Spinning Technology*, Woodhead Publishing Limited, Cambridge, England, 1994.